Text Comprehension and Learning from Text

Bernadette van Hout-Wolters
University of Amsterdam, The Netherlands

Wolfgang Schnotz
University of Tübingen, Germany

(editors)

Taylor & Francis
Taylor & Francis Group

LONDON AND NEW YORK

Library of Congress Cataloging-in-Publication Data

Text comprehension and learning from text / [edited by] Bernadette van Hout-Wolters,
    Wolfgang Schnotz.
        p.   cm.
    Includes bibliographical references.
    ISBN 90 265 1283 X
    1. Textbooks--Readability.   2. Learning, Psychology of. I. Hout-Wolters, Bernadette
H.A.M. van.   II. Schnotz, Wolfgang.
    LB3045. T49 1992
    370.15'23--dc20   92-34188
                                                                               CIP

CIP-gegevens Koninklijke Bibliotheek, Den Haag

Text

Text comprehension and learning from text / Bernadette van Hout-Wolters, Wolfgang Schnotz
ed. – Amsterdam [etc]: Taylor & Francis
ISBN 90-265-1283-X
NUGI 724
Trefw.: tekstwetenschap.

Published by Taylor & Francis
2 Park Square, Milton Park, Abingdon, Oxon, OX14 4RN
270 Madison Ave, New York NY 10016

Transferred to Digital Printing 2007

Lay out: Jan van den Berg, Goirle
Cover design: Rob Molthoff, Amsterdam

ISBN 90 265 1283 X
NUGI 724

Publisher's Note
The publisher has gone to great lengths to ensure the quality of this
reprint but points out that some imperfections in the original
may be apparent

Acknowledgments

Without the assistance of several specialists we would not have been able to finish all the work necessary for the publication of this book.

We would like to acknowledge the great efforts of the following reviewers of the chapters:
Cor Aarnoutse, University of Nijmegen, The Netherlands,
Marianne Elshout-Mohr, University of Amsterdam, The Netherlands,
Joan Peeck, University of Utrecht, The Netherlands,
Ulrich Schiefele, University of Munich, Germany,
Bernd Weidenmann, University of Munich, Germany.

We are also very grateful to Jan van den Berg for his linguistic editing and his lay-out of the manuscript.

Bernadette van Hout-Wolters
Amsterdam, The Netherlands

Wolfgang Schnotz
Tübingen, Germany

Contents

PART II: LEARNER ABILITIES AND ATTITUDES

PART III: PROCESSING AND LEARNING STRATEGIES

1

Text comprehension and learning from text: An introduction

Bernadette Van Hout-Wolters
University of Amsterdam (The Netherlands)
Wolfgang Schnotz
University of Tübingen (Federal Republic of Germany)

In the past years the domain of text comprehension has been an object of intensive research endeavours. Although that development has been reflected by a great number of publications, the situation in Europe is characterised by the fact that researchers from different countries work, relatively independently from one another, on common problems without being sufficiently informed about their colleagues' activities. A reaction to that situation is the establishment of the Special Interest Group "Text Processing"[1] within the European Association for Research on Learning and Instruction (EARLI). The objective of the Special Interest Group is to intensify scientific exchange of theoretical concepts and empirical results, to support an increasing integration of existing approaches, and to show perspectives for a more intertwined international co-operation. This book is one of the results of this co-operation.

The contributions in this book are concentrated, with differing emphases, on three major topics. First, they deal with the significance of different text structures and its cognitive processing in learning from texts. Secondly, they treat the effect of learning abilities and attitudes

---

[1] In 1991 the name of this group was reformulated and became "Comprehension of Verbal and Pictorial Information".

for learning from texts. Finally, they focus on the significance of processing and learning strategies for text comprehension (or knowledge acquisition from texts).

The section on *text structures* is initiated by Grabowski who inquires after the issue of the significance of propositional structure analyses for research on text comprehension and learning from texts. On the basis of theoretical analyses and empirical evidence, he concludes that propositional structure analyses are hardly appropriate for longer expository texts and that they are of limited use, especially concerning instructional psychological research. He considers that the applicability of those approaches is restricted to relatively short narrative texts, the text surface structure of which has a relatively strong effect on comprehension processes.

It is well-known that text structures do not *per se* influence the results of comprehension and learning processes, but only in as far as certain cognitive activities are initiated by them on the learners' part. Therefore, the issue is quite important whether and in how far such structures are discerned by the learners. In his contribution, Vidal-Abarca is dealing with that issue. On the one hand, he reveals that the relevant macroprocesses necessary to detect supraordinate text structures are, at first, based on elementary decoding processes in text understanding; however, later on, they become relatively independent of those decoding processes. On the other hand, his results show that the skill to detect such structures is less important in the processing of explicit main ideas, but very much so in the processing of implicit main ideas. Furthermore, it is shown that the effects of recognising text structures are different according to the level of importance of the text information.

Traditional texts possess a hierarchical-sequential structure, that is, they are linearly organised and consist of text segments which are interlocked so that different levels of hierarchy can be distinguished. At the moment those classical features of texts are questioned in so-called hypertexts, which are based on the use of new information and communication technologies and whereby the learner is enabled to access new information within a non-linearly organised text in a relatively flexible way according to his or her learning prerequisites. The contribution of Samarapungavan and Beishuizen deals with that category of issues. They argue that, in planning and designing such hypermedia, the objective of the teaching/learning processes, the conceptual structure of the respective domain, the structure of presentation as well as the learner's prior knowledge and cognitive abilities are to be taken into account. They further state that, according to those specific conditions, certain contents could more or less appropriately be imparted by hypermedia. In their own study Samarapungavan and Beishuizen reveal that novices are able to reach an increased understanding of a new complex domain by means of a well-structured, indexed hypertext, but they also show that that form of knowledge acquisition is linked to a considerable additional cognitive load, which can, especially for low-ability readers, become a problem. Therefore, the aim of research into that area should concern the development and evaluation of a theoretical model which explains and predicts the reason why or why not non-linear text structures within a hypermedium efficiently support learning processes.

Apart from propositional macro and microstructures, surface structures (including typographic devices connected to them) are of importance. Boscolo, Cisotto and Lucca are concerned with the effect of boxes within texts for primary school children, by means of which signalling of certain chunks of information can occur. Their results show that an effect of that type of signalling on comprehension and recall cannot simply be presupposed. However, the effectiveness of such boxes seems, on the one hand, to depend on the type of information emphasised by them, and, on the other hand, on the learner's comprehension ability level. Additionally, the authors assume that the embedding of the signalled content within the top level structure of the text is quite significant. Thus one could assume that with a normal emphasis, whereby the structurally central text information is additionally highlighted by means of a box, learning is facilitated but also minimally for better learners; however, readers with an average level have a great advantage from that type of signalling. In that context, Boscolo, Cisotto and Lucca pose a number of relevant research questions and call for a theory of text comprehension which systematically relates graphical-signalling processing to the semantic processing of texts.

Research studies on the significance of text structures for comprehension and learning could bear instructional consequences in as far as, in producing and designing texts, structures are realised which promise to especially enhance learning under given conditions. In their contribution, Woudstra and Terlouw describe, from the perspective of Instructional Design, a procedure for drawing up texts, which assumes different text functions and which is based on a process theory of text design. Thereby both the metacognitive skills of the text producer will be improved and the planning and writing activity efficiently supported. In their own empirical studies, Woudstra and Terlouw show that the quality of texts produced in the course can be significantly improved by the systematic reflection about certain design activities.

In the section on *learning abilities and attitudes*, Rimmershaw is, first of all, concerned with the significance of pragmatic knowledge about rhetorical structures, apart from the content knowledge about a domain, especially concerning more difficult texts. She presents the results of an empirical study according to which that pragmatic knowledge does not affect the extent of the learned content, but it correlates with the learner's capacity to grasp the rhetorical structure of the text studied. Thus knowledge of rhetorical structures helps to make especially unstated rhetorical relations explicit and to emphasise unsignalled rhetorical links. However, it could not be corroborated that pragmatic knowledge about rhetorical structures is more important in less well-known domains (or with little prior knowledge) than in well-known domains.

Spiel investigates the issue in how far the capacity to grasp a story, which demands, apart from the identification of the respective formal presentation structure, the understanding of social relations (or conflicts), is age-dependent. By means of an empirical study, she reveals that the ability to recognize interpersonal conflicts within a story does not develop linearly but by leaps within the personal development and that the development of that ability correlates highly with the ability of conditional reasoning. The evidence corroborates the Piagetian hypothesis which postulates a decrease of egocentrism to be a condition for understanding interpersonal conflicts within cognitive development.

Reading and understanding of texts ensues within certain contexts and are supported by educational institutions to a different degree and in differing ways. The contribution of van den Bergh and Rijlaarsdam deals with that topic. The authors ask whether and in how far schools as educational institutions efficiently influence the relation between learners' attitudes towards reading and actual reading behaviour. According to their results, differences among schools have no effect on the relation between affective and behaviourial components of reading, out of which the relatively sceptical view results that measures of school promotion, at least under the present conditions, do not pay off concerning the support of reading activities. However, there are differences among different proficiency levels concerning the relations between reading behaviour and reading performance, whereby one has to presuppose that the type of books read is different according to the different proficiency levels. In general, the relations between attitudes towards reading, reading behaviour, and reading performance could, essentially, be explained by means of individual differences.

In the section on *processing and learning strategies* we see that the process of knowledge acquisition from texts can be fostered by supporting and enhancing appropriate learning activities. Lumbelli concentrates on the effects the feeling of insecurity connected to the evaluation of one's own comprehension has on learning from texts. She reaches the remarkable conclusion that both recognising and ignoring comprehension deficits may be of advantage in knowledge acquisition. According to the results of her study, recognition of one's own lack of comprehension, which is connected to corresponding comprehension standards, offers the possibility to look for additional information and help.

However, if such additional information and help is lacking, that insecurity of comprehension obviously leads to disrupted recording of information in the long-term memory. On the other hand, readers who do not properly understand a text without noticing this and suffer from the disadvantage that they could not use available additional information and aids, have an advantage concerning information processing. This is explained by the fact that they better remember the mental representation constructed by themselves than readers who are conscious of their inadequate text comprehension. Out of these results, Lumbelli deduces instructional consequences both regarding the necessary information offer and necessary aids for metacognitive control of comprehension processes.

The support of knowledge acquisition from texts by means of activation of corresponding processing strategies is the central issue of the next two contributions. Aventissian-Pagoropoulou investigates the effects of cognitive strategy activators, by means of which elaborative and reductive processes could be initiated, on the comprehension and learning performance of experimental subjects. According to her results, especially learners with average capacities benefit from such strategy activation. The interpretation could be that, on the one hand, a specific activation of strategies is no longer necessary for learners with a higher level of capacities. On the other hand, learners with a lower level of capacities should at least be enabled to use such helps, if they were to facilitate learning. In their contribution, García Madruga, Martín Cordero, Luque and Santamaria analyse the effect of a training program for imparting reductive strategies in combination with structure helps in form of advance outlines. The results of their studies show a complex interaction of those different learning aids. The specific effectiveness of the training programme along with simultaneously presented advance outlines corresponds with the assumption that imparting processing strategies frequently serves especially the purpose of using available learning aids in the text effectively.

Although text comprehension ensues frequently within social contexts, research has for a long time mainly concentrated on comprehension processes of an individual working alone. Only recently, that limitation of the research domain has been abolished by works such as those, for example, about reciprocal teaching by Palincsar and Brown. The contribution of Albanese and Antoniotti also belongs to that novel, open line of research, wherein the focus is on how simple monologue-oriented story telling differs from dialogue-oriented story telling, wherein the story-teller addresses questions to the listeners and, during certain periods, leads a dialogue with them. By means of an investigation with preschool children, the authors show that, with dialogue-oriented story telling, the listeners reach an improved, more comprehensive understanding of the contents and produce more elaborated answers to questions than in monologue-oriented story telling. The advantage of dialogue-oriented story telling especially shows in reproduction of information from structure categories which have proved to be especially difficult in Story Grammar research.

The studies presented here are representative for European research on text comprehension and learning with texts, which is not only directed at reaching insights about basic interrelations in the domain, but also at transferring those insights into instructional practice. Therefore, it is important to analyse the research results as to how processes of knowledge acquisition can efficiently be facilitated by means of, on the one hand, appropriate design of text structures, and, on the other hand, imparting and activating appropriate comprehension and learning strategies, taking the respective cognitive and affective learning prerequisites into account. Future research on text comprehension and learning with texts will increasingly face up to such issues of instructional design and instructional practice.

Part I

Text Structures

2

Expository text and propositional text processing

Joachim Grabowski

University of Mannheim (Federal Republic of Germany)

Abstract

The topic of this paper is whether it is worth applying the propositional approach to text processing and readability to the improvement of longer expository texts.

Two versions of a biology text were constructed: a coherent one would, according to the propositional approach, lead to better learning outcomes, than a non-coherent one. To check for instructional alternatives, a reward for good learning results was either announced or not. In a field experiment, 167 students at school read one text under one condition and had to give a written reproduction immediately after reading and again after two weeks.

Reproductions were merely weakly influenced by the texts' propositional structure; visible effects disappeared after two weeks. In contrast, reward caused strong effects, which even became stronger in time.
Finally, some reasons are presented, which account for the fact that previous results within the propositional approach might not be valid for longer expository texts.

Introduction: Text processing and classroom application

The concern of this paper is to examine the use of the propositional approach to text processing for the improvement of expository texts to facilitate learning at school. In this introductory section, we will discuss which characteristics should fit a practicable theory of text processing. After that, the propositional approach will briefly be described. We will present some arguments why this approach, although it dates back about ten years and has already found some successors, could be fruitful yet.

Constraints in practice

With respect to applicability at school, there is a problem with theories in cognitive psychology. For about ten years, the focus of theories of text processing has been on text-reader interactions (Ballstaedt & Mandl, 1988). The results of comprehension and the attendant processes are explained by interactions between specific features of texts and individual characteristics of readers. This approach has led to important progress within cognitive psychology, for example in drawing inferences (Rickheit & Strohner, 1985), making use of previous knowledge (Kintsch, 1988), or constructing mental models (Johnson-Laird, 1983). With regard to a practical use, however, some constraints arise. Interactive teaching programs running on a computer, or facilities to make self-guided learning possible, are not available everywhere for everyday instruction. Expository texts in school books are still a quite important means. But for such texts, the aspect of interaction between text and reader has to be limited to a general model of readers. A good instructional text should work well for every student; there are no different teaching materials for good and not so good students. At the most, a general model of readers might be restricted to a specific group of readers, e.g., 12-year-old pupils. Individual knowledge, individual strategies of information processing and so on must be disregarded.

A second problem still exists with respect to theories in cognitive psychology. It is difficult to make use of models of macrostructure formation (Van Dijk, 1980), schema-driven text processing (Rumelhart, 1975), and mental-model construction (Sanford & Garrod, 1981), when trying to improve expository texts. There is no strict schema for expository texts, as there may be for narratives (Thorndyke, 1977), which might control top-down processes. On the other hand, a schema depending on semantic content, or a mental model, must first be developed by reading the text and therefore cannot be assumed to be a prerequisite of the comprehension process.

The propositional approach

The constraints set out in the previous paragraph are in our opinion best fitted by the propositional approach, represented by the cyclic model of text processing (Kintsch & Van Dijk, 1978). It includes a conception of text comprehensibility introduced by Kintsch & Vipond (1979). We will briefly describe this approach.

Propositions can be considered to be basic elements of a knowledge structure (Kintsch, 1974; a linguistic predecessor was Fillmore, 1968). They are conceptualised by connections between concepts and they consist of a predicate and one or more arguments; the predicate specifies the relation between arguments. There are some manuals for propositional analysis (Bovair & Kieras, 1981; Sowarka et al., 1983; Turner & Greene, 1977), which try to enable researchers to represent a text's semantic content. With respect to mental representation, the "psychological relevance" of propositions has been confirmed in a lot of studies (e.g. Beyer, 1986; Kintsch & Keenan, 1973; Ratcliff & McCoon, 1978; Yekovich & Manelis, 1980). A basic assumption of the Kintsch & Van Dijk model of text processing is that a reader copies some 20 propositions, which are extracted from the text by subsemantic reading processes, into his or her working-memory. Working-memory is characterised by its limited size. Then it is tried to establish coherence between these propositions by constructing a hierarchical graph of connected propositions. The high-level propositions are kept in the short-term buffer, which is part of the working-memory, while the graph so far constructed is copied into the long-term memory. The next portion of textual propositions is read, and the same process is starting again and so on. Text processing is facilitated when coherence can easily be established, i.e., without the necessity of drawing inferences, reconstructions of a graph once built up, or reinstatements of formerly processed propositions, already stored in long-term memory. These kinds of operations are regarded as strongly demanding the reader's cognitive resources. By means of this cyclic process the so-called levels effect is also explained; i.e., the higher probability of propositions to be recalled when they are positioned at a high level within the hierarchical graph of coherence, the more often a proposition is kept in short-term buffer and runs a processing cycle, the better it will be stored and the higher its probability of being recalled will be.

The propositional model does fit the characteristics for general use, which have been set out above, although it is a theory of text-reader interaction, too. The readers' component is basically the size of the working-memory's short-term buffer. But, as we will show in section 2 below, the propositional structure of a text, that may lead to better processing, can be improved without depending on a specific size of a reader's working-memory.

The problem of economy

Although some critical points have been stated against the propositional approach (Christmann, 1989; Grabowski, 1991; Schnotz, 1988), for instance concerning the problem of referential identity of concepts within a text, for the most cases it has been summed up as being objective and robust; therefore, it can be seen as a useful working tool for the handling of texts in cognitive and instructional psychology.

But for longer expository texts, there is a problem with the method's economy. Propositional research, which has mainly been carried out from the late sixties to the eighties, is predominantly done with short texts, mostly with narratives, and preferentially in the lab. There are only a few exceptions, in which the propositional approach has been scrutinised for longer texts, especially in an area of application. Mostly, there are a-priori statements, suggesting that a propositional analysis would be too costly for longer and more complex texts. So the question whether to apply the propositional approach to longer texts or not, was seen as a question of economy (and deferred). Indeed, the procedure requires a lot of time in every way. But, in our opinion, a method's economy can be assessed only as a relation between cost and outcome. The fact that a certain method is costly to apply is no justification for totally giving up the method's evaluation. For the motor-car industry, for example, the development of a new series of car, of course, takes several years. That obviously does not reduce economy, because the profit will cover the costs of investigation. Expository texts in school books are printed in some ten thousand copies. Therefore, we think that the costs for propositional control of textual structure would be justified and balanced out if by that means improvement for some relevant criteria of understanding, such as recall, remembrance or drawing inferences, would be achieved. Another aspect of economy, however, is that the same amount of improvement should not be reached in another way more easily.

So our aim was to provide an answer to the following questions: (1) Does a better propositional (i.e., coherential) structure of a longer expository text lead to better learning outcomes at school? (2) Is the possible improvement worth the effort of controlling texts propositionally?

A field experiment

The following field experiment was conducted: two versions of a text about the neurophysiology of sleeping in human beings were constructed. They differed in their propositional structure of coherence. According to the propositional model of text processing, for one textual version it was easy to establish coherence ("coherent text"). This should make text processing easier and learning outcomes better. For the other textual version ("non-coherent text"), it was expected that processing would be very difficult, what would subsequently lead to poorer results. All other characteristics were tried to be kept constant for both texts (see Table 1).

According to the Kintsch & Van Dijk model, text processing can be simulated (see Table 2). It can be seen that for each presumed capacity of the short-term buffer (which may slightly vary from reader to reader within a certain range) the coherent text (on the left side of the table) is easier to process than the other one. This is how we tried to ensure the aspect of generality even in a model of text-reader interaction: despite individual characteristics of readers (here: concerning their short-term buffer capacity), the non-coherent text is assumed to be harder to process than the other one.

Table 1: Characteristics of experimental texts

characteristic	coherent text	non-coherent text
number of sentences	36.0	35
number of words	454.0	447
number of different words	269.0	271
Type-Token-Ratio	0.59	0.61
number of syllables	861.0	858
words per sentence	12.61	12.78
number of propositions	204.0	197
propositions per sentence	5.67	5.63
arguments per proposition	1.91	1.90
words per proposition	2.23	2.27

To fully analyse a longer text propositionally and to construct its graph of coherence – as far as possible in the non-coherent case – is, of course, very costly and time-consuming; the results in terms of learning improvement by propositional control of a text must therefore be pretty good to balance economy.

The second independent variable in the experiment described here was that each text version was presented with or without the announcement of a reward for good results (maximally DM 10). This was to check the aspect of propositional economy, viz. that some other manipulation could more easily lead to the same improvements in text processing than a propositional control of texts. It was expected that, for all participants, the announcement of a reward would influence important reader variables (e.g., attention, effort) into a positive direction, independent of individual ability (for a detailed explanation how possible effects of rewards could be reconstructed theoretically within the propositional approach, cf. Grabowski, 1991; Heckhausen, 1989).

Table 2: *"Behaviour" of experimental texts within the cyclic model of text processing (Kintsch & Van Dijk, 1978) (+ = automatic generation of coherence; K = break in coherence)*

Capacity of working-memory's short-term buffer (number of propositions)

Input	WM	coherent text					non-coherent text				
		3	4	5	6	7	3	4	5	6	7
Sent.	1	+	+	+	+	+	+	+	+	+	+
Sent.	2	+	+	+	+	+	+	+	+	+	+
Sent.	3	+	+	+	+	+	K	+	+	+	+
Sent.	4	+	+	+	+	+	K	K	K	K	K
Sent.	5	K	+	+	+	+	+	+	+	+	+
Sent.	6	K	+	+	+	+	K	K	K	K	K
Sent.	7	+	+	+	+	+	+	+	+	+	+
Sent.	8	+	+	+	+	+	+	+	+	+	+
Sent.	9	+	+	+	+	+	K	K	K	K	K
Sent.	10	+	+	+	+	+	K	K	K	K	K
Sent.	11	+	+	+	+	+	K	K	K	K	K
Sent.	12	+	+	+	+	+	+	+	+	+	+
Sent.	13	+	+	+	+	+	K	K	K	+	+
Sent.	14	+	+	+	+	+	K	K	K	K	K
Sent.	15	+	+	+	+	+	+	+	+	+	+
Sent.	16	+	+	+	+	+	+	+	+	+	+
Sent.	17	+	+	+	+	+	K	K	K	K	K
Sent.	18	+	+	+	+	+	K	+	+	+	+
Sent.	19	K	K	+	+	+	+	+	+	+	+
Sent.	20	K	K	K	K	+	+	+	+	+	+
Sent.	21	+	+	+	+	+	K	K	K	K	K
Sent.	22	+	+	+	+	+	K	K	K	K	K
Sent.	23	+	+	+	+	+	+	+	+	+	+
Sent.	24	+	+	+	+	+	+	+	+	+	+
Sent.	25	+	+	+	+	+	+	+	+	+	+
Sent.	26	+	+	+	+	+	K	K	K	K	K
Sent.	27	+	+	+	+	+	K	K	K	K	K
Sent.	28	+	+	+	+	+	K	K	K	K	K
Sent.	29	+	+	+	+	+	+	+	+	+	+
Sent.	30	+	+	+	+	+	+	+	+	+	+
Sent.	31	+	+	+	+	+	K	K	K	K	K
Sent.	32	+	+	+	+	+	+	+	+	+	+
Sent.	33	K	K	K	+	+	+	+	+	+	+
Sent.	34	+	+	+	+	+	K	K	K	K	K
Sent.	35	+	+	+	+	+	+	+	+	+	+
Sent.	36	K	K	K	+	+					
Number of breaks		6	4	3	1	–	17	15	15	14	14

We know from locus-of-control theories that students with lower capabilities might make no efforts, even when a reward is announced, because they do not see any opportunity for an improvement of their results (Meyer, 1973; Weiner *et al.*, 1972). This could lead to paradoxical effects of rewards. To avoid this, reward was announced collectively, that is, that every participant would be rewarded, if the average result of a whole experimental group was good enough. Under another condition, we additionally varied the interestingness of expository texts. We report about these aspect elsewhere (cf. Grabowski, 1991).

Within the experimental design of two factors with two levels each, there were 167 subjects at the age of about sixteen, who stayed in the school classes they belonged to. The classes were at high-school level (the German "Gymnasium"). Each student participated in one condition during a normal weekly lesson in biology. Subjects were instructed to read the respective text as if they were preparing themselves for a class test. Reading time was limited to 8 minutes, which was long enough for everybody to read the text at least twice. Students had to give a complete written reproduction immediately after reading and after two weeks again.

Possible effects between schools (and between classes respectively), which might be based on other factors than experimental manipulation, were controlled by a covariate.

If the announcement of reward would lead to better results, this could depend on more efforts during reproduction, or on more intensive retrieval, etcetera. To ensure that reward really affects the processing and comprehension of a text, two control groups were introduced. Whereas for the experimental groups reward was announced before reading the (coherent or non-coherent) text, the students of the control groups first read the (coherent or non-coherent) text without knowing anything about a reward, after which the reward was announced, and then subjects were asked for their written reproductions. The experimental design is illustrated in Table 3.

Table 3: Experimental design (number of subjects)

text	with reward	without reward	reward control
coherent	38	27	13
non-coherent	42	32	15

We expected two main effects: better learning with the coherent text than with the non-coherent one, and better learning if a reward was announced than if it was not. Our decision about economy of the propositional approach depends on the strength of these effects as well as on the occurrence of a first-order interaction.

Results

The written reproductions were analysed in the following manner. *N of words* is a simple measurement of the total amount of reproduction. *N of units* is the number of correctly reproduced semantic units, out of the 63 units into which both texts were divided. This second variable is called *n of units* instead of *n of propositions*, because it was taken into consideration that not every single proposition can be reproduced independently of any other (see section 4 below). If a unit was reproduced incorrectly or in a wrong context, it was counted as an *error*. *N of concepts* indicates how many of thirteen technical terms, occurring in both texts, were reproduced. *N of items* is the number of correct solutions for ten multiple-choice items, which could not be solved directly from the text, but required drawing inferences. These items were presented after the second reproduction only. *Rank order* refers to the rank correlation between the sequence of semantic units in the original text and the sequence in the reproduction.

The results of the control group can briefly be summed up: when a reward was announced *after* the text had been read, subjects behaved in every respect as if no reward had been announced; this is true for both the coherent and the non-coherent text. Therefore, any effect of a reward announced *before* reading will indeed depend on influencing text processing.

For the experimental groups, analyses of variance were computed; the results are summarised in Table 4. All significant differences between means were in the expected direction, i.e., if there is an effect, better scores occurred for the coherent text and the reward condition, respectively. There were no interactions between the two independent variables.

For variables measuring the amount of reproduction (n of words, n of units), there was an effect of propositional coherence on the immediate reproduction. These effects disappeared after two weeks. The opposite is true for effects of reward. Under all conditions, there were only few errors; for the immediate reproduction, readers of the coherent version produced significantly less errors than readers of the non-coherent text. Rewarded subjects retrieved more technical terms within their second reproduction after two weeks (n of concepts) and performed better in drawing correct inferences (n of items). There were no differences between average rank coefficients. That indicates that the reproductions of the non-coherent text were sequenced in the same way as the non-coherent text itself, whereas the reproductions of the coherent version were arranged like the coherent original.

Table 4: *Summary of results (* = sign. with p < 5%; ** = sign. with p < 1%; Eta² = explained variance); no interaction for each dependent variable. 1 = immediate test; 2 = test after two weeks.*

Effects: variable	coherence	reward	Eta²
n of words 1	*	**	11%
n of words 2	n.s.	**	24%
n of units 1	**	n.s.	9%
n of units 2	n.s.	**	16%
n errors 1	**	n.s.	8%
n errors 2	n.s.	n.s.	2%
n concepts 1	n.s.	n.s.	8%
n concepts 2	n.s.	*	5%
n of items 2	n.s.	**	4%
rank of order 1	n.s.	n.s.	8%
rank of order 2	n.s.	n.s.	6%

In Table 4, the two times of reproduction collection were treated as separate variables. What happens from immediate reproduction towards a reproduction after two weeks becomes more clear in Table 5, where the two times are treated as repeated measurements. Of course, all reproductions become worse. If there were any differences between the textual versions for the immediate reproduction, they disappeared after two weeks. On the other hand, effects of reward became stronger in time.

Additionally, we tried to confirm the levels effect for both times of reproduction. But we found that a semantic units' probability of recall is not even affected by the level it takes within the hierarchical graph of coherence.

Table 5: Analyses of repeated measurements (effects with $p < 5\%$)

effects: variable	time	coherence x time	reward x time
n of words	decreases	--	more distant
n of units	decreases	closer	more distant
errors	--	closer	--
concepts	decreases	--	--
rank of order	decreases	--	--

Discussion

To summarise, there are only weak effects of a propositional control of textual coherence. If any effects arise, they disappear after two weeks. So a propositional control of expository texts does not seem to influence learning and remembering as strongly as expected. In contrast, the announcement of a reward improves learning outcomes much more; after two weeks the effects of reward become stronger. Considering that instruction at school should for the most part be learning for the future (and not for the class test tomorrow), a propositional control of expository texts seems not to be very fruitful, especially with respect to the efforts needed for analysing longer texts propositionally.

Therefore, the final conclusion is that it is indeed not economical and not successful to have a propositional control for longer expository texts. How can it be explained that there are only weak effects? Not even the levels effect could be confirmed, whereas the state of research shows stable effects, most recently in the work of Beyer in 1987. In my opinion there are at least four critical points, which in part have been referred to by other authors, too (cf. Christmann, 1989; Schnotz, 1988). These points are set out briefly for further discussion:

(1) Research has mostly been done with narratives. In these cases, the propositional structure of hierarchy in a text is not independent of the logical structure of a plot. If a story is about a hunter who does something and who additionally has a green jacket on, the proposition containing the hunter's action will probably be put on a high hierarchical level. This is because the action, the action's patient, etcetera, will be good high-level organisers for everything that will happen later. The proposition containing the attribution of his jacket's colour will be subordinated. Within empiry, the story might be reproduced without remembering the jacket's colour. But the jacket's colour will never be reproduced without remembering the story's action. The result is a higher probability of reproduction for the high-level propositions (containing actions), which, in this line of thoughts, is not necessarily caused by the hierarchical structure but by logic. Consequently, not every result obtained from narratives might be valid for expository texts.

(2) Apparently, results may not be transferable from short texts to longer texts. When experimenting with short texts, one can never be sure whether the subjects' reproductions are based on linguistic-episodic remembrance (concerning the text's linguistic surface) or whether they are really based on learning (concerning the text's meaning). The closer reproductions are to a text's surface, the better a text's propositional and hierarchical structure will predict the results of reproduction.

(3) Propositions are descriptions of connections between concepts. They do not, however, reflect the structure within a concept. An example: a unit in the experimental texts was that it is not necessary to shave the hair of the patient's head for an electroencephalogram. This unit was reproduced very often, although it was not important for the overall

structure, and although it was placed at the lowest level of hierarchy within the graph of coherence. The theory of mental-model construction will lead onward because of this fact, of course. If we have a look at the internal organisation of concepts, one would assume that not the abstract-semantic modality of representation is the salient one, but motor components and sensorial images (Engelkamp 1990; Engelkamp & Zimmer, 1983; Herrmann, 1985). The propositional approach deals with semantic components of concepts only.

(4) Our final point is that reproductions are usually analysed in such a way that the number of correctly reproduced propositions is ascertained. This measurement only makes sense if it is possible to reproduce each proposition independently of any other. But this is not true for negation and for embedded propositions. It would have been better to work with conditional probabilities of reproduction, but this aspect has not been considered within the simulations of text processing introduced by the Kintsch group. We assume that, for many cases, researchers have had to try to reach their goal (and may have succeeded) by a more or less modified or even hand-made way of propositional analysis of reproductions, which is, unfortunately, for the most part not described in detail in research reports.

Propositional analysis was and is an important tool for handling a text's meaning, not in the least by its stimulating effects on respective research since the publication of Kintsch & Van Dijk (1978). But its use for research closely related to applicational problems under applicational conditions, is obviously limited. A theoretical integration of semantic, motivational and holistic aspects of text processing may lead onward, as soon as it provides adequate methods for the analysis of instructional material and its cognitive (and perhaps emotional-motivational) processing, which are, outside the laboratory, as robust, clear and operative as the propositional approach has been inside.

32

References

Ballstaedt, S., & Mandl, H. (1988). The assessment of comprehensibility. In U. Ammon, N. Dittmar & K. Mattheier (Eds.), *Sociolinguistics: An international Handbook of the Science of Language and Society* (2. Halbband) (pp. 1039-1052). Berlin: De Gruyter.

Beyer, R. (1986). Investigations on text processing with regard to the model of Kintsch and Van Dijk (1978). In F. Klix & H. Hagendorf (Eds.), *Human memory and cognitive capabilities. Mechanisms and performances* (pp. 871-885). Amsterdam: Elsevier.

Beyer, R. (1987). Psychologische Untersuchungen zur Textverarbeitung unter besonderer Berücksichtigung des Modells von Kintsch und Van Dijk (1978). *Zeitschrift für Psychologie mit Zeitschrift für Angewandte Psychologie, Suppl. 8*, 1-80.

Bovair, S., & Kieras, D. (1981). *A guide to propositional analysis for research on technical prose* (Technical Report No. 8). University of Arizona.

Christmann, U. (1989). *Modelle der Textverarbeitung: Textbeschreibung als Textverstehen.* Münster: Aschendorff.

Engelkamp, J. (1990). *Das menschliche Gedächtnis.* Göttingen: Hogrefe.

Engelkamp, J., & Zimmer, H. (1983). *Dynamic aspects of language processing.* Berlin: Springer.

Fillmore, C. (1968). The case for case. In E. Beach & R. Harms (Eds.), *Universals of linguistic theory* (pp. 1-88). New York: Holt, Rinehart & Winston.

Grabowski, J. (1991). *Der propositionale Ansatz der Textverständlichkeit: Kohärenz, Interessantheit und Behalten.* Münster: Aschendorff.

Heckhausen, H. (1989). *Motivation und Handeln* (2. völlig überarb. und erg. Aufl.). Berlin: Springer.

Herrmann, T. (1985). *Allgemeine Sprachpsychologie.* München: Urban & Schwarzenberg.

Johnson-Laird, P. (1983). *Mental Models: Towards a cognitive science of language, inference and consciousness.* Cambridge: Cambridge University Press.

Kintsch, W. (1974). *The representation of meaning in memory.* Hillsdale, NJ: Lawrence Erlbaum Associates.

Kintsch, W. (1988). The role of knowledge in discourse comprehension: A construction-integration model. *Psychological Review, 95*, 163-182.

Kintsch, W., & Keenan, J. (1973). Reading rate and retention as a function of the number of propositions in the base structure of sentences. *Cognitive Psychology, 5*, 257-274.

Kintsch, W., & Van Dijk, T. (1978). Toward a model of text comprehension and production. *Psychological Review, 85*, 363-394.

Kintsch, W., & Vipond, D. (1979). Reading comprehension and readability in educational practice and psycholgical theory. In L. Nilsson (Ed.), *Perspectives on memory research* (pp. 329-365). Hillsdale, NJ: Lawrence Erlbaum Associates.

Meyer, W. (1973). *Leistungsmotiv und Ursachenerklärung von Erfolg und Mißerfolg.* Stuttgart: Klett.

Ratcliff, R., & McKoon, G. (1978). Priming the item recognition: Evidence for the propositional structure of sentences. *Journal of Verbal Learning and Verbal Behavior, 17*, 403-418.

Rickheit, G., & Strohner, H. (1985). *Inferences in text processing.* Amsterdam: North Holland.

Rumelhart, D. (1975). Notes on a schema for stories. In D. Bobrow & A. Collins (Eds.), *Representation and Understanding* (pp. 211-236). New York: Academic Press.

Sanford, A., & Garrod, S. (1981). *Understanding written language: Explorations of comprehension beyond the sentence.* New York: Wiley.

Schnotz, W. (1988). Textverstehen als Aufbau mentaler Modelle. In H. Mandl & H. Spada (Eds.), *Wissenspsychologie* (pp. 299-330). München: Psychologie Verlags Union.

Sowarka, B.Ü., Abel, U., & Michel, J. (1983). *Menschliche Textverarbeitung und propositionale Analyse. Eine Anleitung zur propositionalen Darstellung von Texten* (Arbeitspapiere zur Linguistik, 18). Berlin: Techn. Universität, Institut für Linguistik.

Thorndyke, P. (1977). Cognitive structures in comprehension and memory of narrative discourse. *Cognitive Psychology, 9,* 77-110.

Turner, A., & Greene, E. (1977). *Constructions and use of a propositional text base* (Technical Report No. 63). University of Colorado, Institute for the Study of Intellectual Behavior.

Van Dijk, T. (1980). *Macrostructures. An interdisciplinary study of global structures in discourse, interaction, and cognition.* Hillsdale, NJ: Lawrence Erlbaum Associates.

Weiner, B., Heckhausen, H., Meyer, W., & Cook, R. (1972). Causal ascriptions and achievement behavior: A conceptual analysis of effort and reanalysis of locus of control. *Journal of Personality and Social Psychology, 21,* 239-248.

Yekovich, F., & Manelis, L. (1980). Accessing integrated and nonintegrated propositional structures in memory. *Memory & Cognition, 8,* 133-140.

3

Important information in expository prose: The role of awareness of textual structure and of the skill in ranking ideas in a text

Eduardo Vidal-Abarca
University of Valencia (Spain)

Abstract

This research analyses the role that awareness of textual structure and skill in ranking ideas in a text according to their relative importance in the text structure play in comprehension and recall of the most important information in expository prose. We were also interested in comparing the results obtained with different measures of awareness of textual structure. Thirty-two fifth-grade children from an Elementary School of Valencia (Spain) participated in this study. Four independent measures were taken: decoding, intelligence, evaluating sentences in a text according to their relative importance to the textual structure, and three different tests of awareness of textual structure. Three dependent measures were selected: writing down the explicit main idea of several texts, writing down the implicit main idea of other texts, and recalling the most important information of a text. Data analysis confirmed the role that awareness of textual structure plays in getting the implicit main idea and in recalling the most important information of expository prose. Nevertheless, we obtained different results with the different measures we employed. The role of the skill in ranking ideas in a text is confirmed only for getting the implicit main idea. Some theoretical and practical educational implications are discussed.

Introduction

According to Van Dijk and Kintsch's model (1983), schematic super-structures of textual structures play a role in the macrostructure formation in that they assign textual functions to macropropositions. This role is played in comprehension and recall tasks. Two assumptions of this models are relevant for our research. The first is that schematic super-structures must be not only in the text, but also in the reader's mind. The second is that schematic superstructures are learnt.

The first assumption is confirmed by several studies on comprehension and expository prose. These studies demonstrate that subjects who are aware of these textual structures recall more top level information than others less conscious of textual organization (Meyer, Brandt & Bluth, 1980; Taylor & Samuels, 1983). Similarly, the more conscious subjects are, the better they organise their recall and the more effective they are with summary strategies (Sanchez, 1988).

An issue that is faced by all these studies is how to measure the awareness of textual structure. Three kinds of measures can be distinguished (see Figure 1). The first, and most frequently used, method consists of evaluating to what extent subjects' recall is organised in a similar way as the written text (Meyer, Brandt & Bluth, 1980; Brooks & Dansereau, 1983; Meyer & Freedle, 1984; Richgels et al., 1987). We call this method "recall texts". The second method consists of asking subjects to produce several sentences, the main idea, or and entire text according to any kind of textual structure previously suggested (Englert & Thomas, 1987; Englert et al., 1988; Scardamalia & Paris, 1985). We call this "productions tests". The third method, which we call "comprehension tests", consists of matching or grouping texts with the same structure (Richgels et al., 1987; Cook & Mayer, 1988), or rating the "degree of fit" of several sentences with an initial paragraph that signalled a definite textual structure (Englert & Hiebert, 1984). The results obtained with these different kinds of measures cannot be compared because they do not use equivalent text, nor do they depart from the same classification of textual structure, nor are the subjects comparable.

Type of test	Authors	Sub-jects	Textual structure*
RECALL TESTS Degree of correspond-ence between text structure and recall structure • organised • partially organised • not organised	Meyer, Brand & Bluth (1980) Brooks & Dansereau (1983) Taylor & Samuels (1983) Meyer & Freedle (1984) Richgels et al. (1987)	9 Univ. 5,6 Univ. 6	Cpar/ProbSol Cpar Enu Caus/Cpar/Enu/ ProBSol Cpi/Cpar/Cau/ ProBSol
PRODUCTION TESTS Producing • several detailed sen-tences • main idea • complete text according to textual structure previously sug-gested	Englert & Thomas (1987) Englert et al. (1988) Scardamalia & Paris (1985)	3-4,6-7 3,6 4,6,10 Univ.	Des/Enu/Seq/ Cpar Enu/Seq/Cpar Essays (for- against)
COMPREHENSION TESTS • Matching or grouping texts with the same structure • Rating the degree of fit of several sen-tences with an initial paragraph	Richgels et al. (1987) Cook & Mayer (1988) Englert & Hiebert (1984)	6 Univ. 3,6	Cpi/Cpar/ Cau/ProbSol Gen/Enu/Deq/ Cla/Cpar Des/Enu/Seq/ Cpar
* CODE for textual structures: Cpar = comparative/contrast; ProbSol = problem solution; Enu = enumeration; Cpil = compilation; Cau = causation; Des = description; Seq = sequence; Gen = generalisation; Cla = classification			

Figure 1: Examples of research studies employing different measures of textual structure awareness

As was previously stated, Van Dijk and Kintsch's model assumes that schematic superstructures exist not only in the text but also in the readers' mind, which implies a process of learning. This fact has been demonstrated by several studies in which subjects of different age groups have been instructed in all kinds of "structure strategy" (Meyer, 1984). These subjects improved not only their comprehension but also their recall of textual information, especially the top level information

(Brooks & Dansereau, 1983; Armbruster *et al.*, 1987; Cook & Mayer, 1988; Vidal-Abarca, 1990).

The skill in identifying the most important information of a message (which was the second variable we studied), has been signalled as an important metacognitive skill (Brown, 1980). A test of this skill consists of asking subjects to rank text sentences according to their relative importance in the text (Winograd, 1984). Good and poor readers scored differently in this task and the task is strongly related to summarisation tasks.

The first goal of this study was to check the role of the two variables we explained above: awareness of textual structure and the skill in ranking ideas in a text, in comprehension and recall of the most important information in expository prose. We first hypothesised that these two variables would predict the ability of getting the explicit and implicit main idea from expository text; secondly, that both factors would also predict the recall of the most important information; and, thirdly, that reading microprocesses measured by some indicators of decoding would not predict reading macroprocesses. An additional but central purpose of our study was to compare the results obtained with different measures of textual awareness.

Method

Subjects

Thirty-two fifth-grade children from an Elementary School of Valencia (Spain) participated in this study. They were all from middle socio-economic level and there were approximately equal numbers of girls and boys.

Material

Several texts were elaborated to be used in the self-developed measures. They were all adapted from fourth and fifth-grade content area textbooks. The adaptation consisted of increasing their coherence and clarifying their structure following Anderson & Armbruster's (1984) recommendations. Half of them were written to conform as closely as possible to comparison/contrast rhetorical structure and the other half to enumerative structure according to Englert and Hiebert's (1984) definition of rhetorical structures in expository prose. All the texts, except the one we used in recall tests, were between 119 and 207 words long (average of 166.5). The text we used in the recall test was 362 words long, and its content was not included in the school curriculum, namely, the comparison/contrast between "space stations and space shuttles" in four characteristics: nature, use, shape, and size.

Measures

Four independent measures were taken. First, we employed two subtests from Test de Análisis de Lecto-Escritura (T.A.L.E.): Words and Text. Two different decoding measures were taken: sum of errors and sum of reading time. Secondly, we measured intelligence with the test Bateria de Aptitudes Diferenciales y Generales (B.A.D.Y.G., form E). Again two different measures were taken: verbal intelligence and non-verbal intelligence. Thirdly, we measured the skill at ranking sentences according to their relative importance in textual structure. Children were asked to rank five sentences from each of the two texts according to their relative importance in textual structure, rating 2 for high importance, 1 for medium, and 0 for low importance. An item of this self-developed measure, and of some others we will explain below, can be seen in Figure 2.

The fourth type of independent measures assessed the subjects' textual structure awareness. The first task used to test this skill was to match two texts according to their textual structure (enumerative or comparison/contrast) and to explain the reasons for the matching. This was adapted from Richgels *et al.* (1987). The matching task material consisted of a three-page booklet with three passages on each page.

Measure: Ranking sentences in a text.
Text:
There are persons who think that spiders belong to a type of insects, but that is not the case. Spiders belong to a group called "arachnoids". The spiders' body is different from that of insects. Among other things, one of the differences can be observed by counting legs. While all insects have six legs, spiders have eight. Flies or ants, for example, are insects and all of them have six legs. Another difference can be seen by observing how the body of these little animals is divided. All insects have their body divided into three parts: the head, the thorax and the abdomen. However, the spiders' body has only two parts. Besides, almost all insects have wings and a pair of antennas on their heads. Spiders, however, do not have any of these two types of organs. Therefore, spiders cannot fly.

Rank each of these sentences scoring 2 if the sentence is very important in the text, 1 if it is not very important, and 0 if it is hardly important or if it is a detail.
.... There are persons who think that spiders belong to a type of insects.
.... The body of spiders is different from that of the insects.
.... One of the differences can be observed by counting their legs.
.... Flies or ants, for example, are insects and all of them have six legs.
.... Another difference can be seen by observing how the body of these little animals is divided.

Measure: Matching texts.
(Passages were similar to the ones used in the ranking sentences task)
TEXT 1. (Passage with comparative/contrast textual structure)
TEXT 2. (Passage with comparative/contrast textual structure)
TEXT 3. (Passage with enumerative textual structure)
Passage 1 should be matched with Passage   □ "2"   □ "3"   *Why?*
....
....

Measure: Detection of target and distractor sentences.
Which of the sentences fits in with the initial paragraph? Write "YES" if you think that the sentence belongs to the paragraph, "YES?" if you think it sort of belongs to the paragraph, "NO?" if you think it sort of does not belong to the paragraph, and "NO" if you think it definitely does not belong to the paragraph.

Human beings can take various precautions to look after their special senses such as sight and hearing. To avoid damage to our sense of sight, it is convenient to watch T.V. from a certain distance away from the T.V. set.

A. When we read or write, it is not good to look down too close at the paper.
B. Bats have their sense of hearing far more developed than human beings.
C. Some animals, such as the eagle, have much better eye-sight than human beings.
D. To clean our ears we should avoid using hard or fine-pointed objects.

Figure 2: Example of self-developed measure

On each page the first text and either the second or the third passage (the match) had the same text structure. The other passage on the page (the foil) did not have the same text structure as the first passage.

The three passages had different content. The first passage was printed in bold type, and the second and third in plain type. The first page was used to explain the task to children by the instructor who not only modelled the matching but also the reasons for it. The second task was similar to Englert and Hiebert's one. The test had six items, three of them were comparative, and the rest were enumerative. In every item the children were asked to rate the "degree of fit" of four sentences with an initial paragraph choosing among four possibilities, from "the sentence belongs to the paragraph" to "it does not belong to the paragraph". Two were target sentences and the other two were distractors. The initial paragraph consisted of two sentences, a main idea sentence with an evident textual structure and a clear topic, and a detail sentence consistent with the former. In the third task, children were asked to read a text with comparison/contrast structure, after which they had to carry out a task aimed at distracting them; later they were asked to write down everything they could recall from the text; we measured the degree of correspondence between the text structure and the structure recalled by the children.

Three dependent measures were obtained. The first one was to write the main idea explicitly stated in four texts. The second one was to write the implicit main idea of two texts. In both cases, half of the texts were comparative and the other ones were enumerative. The third dependent measure tested the skill in recalling the most important information in a text. The recall task we used was the same as the one described above. The text was divided into idea units (Mayer, 1985), after which they were classified by expert judges into three thirds. Recall of the three most important idea units is the aim of this work, although we also included results of the three least important idea units to emphasise those data.

Procedure

All the measures, except the one for decoding which was taken individually, were collectively administered during regular class periods. Subjects were given as much time as needed to respond, except for the recall task in which they had 10 minutes to read the text and an equal period for free recall after a short distractor task. Subjects were given practice trials appropriate to each task.

Scoring

The scoring system for the measure of ranking sentences was: coincidence of the correct answer with the children's answer, 2 points, one point of difference between both, 1 point, and two points of difference, 0 points. In case of the first measure of textual structure awareness (matching texts) we scored 1 point for each correct matching and structural reason, 0.5 point for a correct matching without and explicit reason, and 0 points for all other responses. The second measure of textual structure awareness (rating the degree of fit of sentences with an initial paragraph) was scored 1 through 4 depending on the accuracy in the rating. Subjects received separate scores for target and distractor sentences. In the third measure of textual structure awareness (degree of correspondence between the text structure and the structure recalled) the scoring system combined two criteria: number of main idea recalled (Taylor & Samuels, 1983), and percentage of recall organised according to text structure (Brooks & Dansereau, 1983). According to these, we classified the recall as "organised" if more than 80% had the same comparative structure as the text and/or contained at least three out of five top level ideas; "partially organised" if 80-50% of the recall had the same comparative structure as the text and/or contained at least two out of five top level ideas; and "not organised" in all other cases.

In case of the first dependent measure we scored 1 if subjects selected the explicit main idea and 0 if they did not it. Getting the implicit main idea was scored 1, 0.5 or 0 if children produced the correct answer, only the theme or topic, and all other responses, respectively. For the third dependent measure (recall of idea units) we scored 1 or 0 depending on their presence or absence in children's recall.

In case of open answer measures, one third of the protocols were independently scored by two judges. Interrater agreement ranged from 0.88 to 0.96. All the test had previously been answered by undergraduate students to ensure the correctness of our criteria about right and wrong answers.

Results

In order to test our hypotheses, we used regression analysis and simple analysis of variance. The first hypothesis stated that awareness of textual structure and the skill in ranking ideas in a text would predict the ability of getting the explicit and implicit main idea from expository text. The results of the multiple regression with the two first dependent variables, getting the explicit and implicit main idea from texts, are presented separately in Tables 1 and 2.

Table 1: *Results obtained in the multiple regression for the sum of explicit main idea from texts (* $p<.05$, ** $p<.01$)*

Variables	Stand.coef.	T	p
Sum of errors in decoding	-0.435	1.327	0.20
Sum of reading time	0.392	1.192	0.24
Verbal intelligence	-0.058	0.138	0.89
Non-verbal intelligence	0.436	1.257	0.22
Ranking sentences	0.150	0.729	0.47
Matching texts	-0.193	0.832	0.41
Detection of target sentences	0.094	0.432	0.67
Detection of distractor sentences	0.388	1.431	0.17
Constant	0.000		

$N=32$; $R^2=0.375$; $F_{reg}=1.777$

The pattern of results is very different. None of the variables contribute significantly to explain the scores in case of explicit main idea, but some of them do so in the second dependent measure, getting the implicit main idea. A measure of awareness of textual structure (matching texts) is significant, and another one (detection of target sentences) is close to the usually admitted level; this can also be said in case of ranking sentences. The other measure of awareness (detection of distractor sentences) is far from the significant level. We also found that verbal intelligence is a highly predictive variable in case of getting the implicit main idea from texts. This result was not hypothesised but it is consistent with other studies.

*Table 2: Results obtained in the multiple regression for the sum of implicit main idea from texts (** p<.01)*

Variables	Stand.coef.	T	p
Sum of errors in decoding	-0.179	0.722	0.477
Sum of reading time	0.462	1.857	0.076
Verbal intelligence	1.074	3.373	0.002**
Non-verbal intelligence	-0.404	1.540	0.137
Ranking sentences	0.310	1.989	0.058
Matching texts	0.488	2.775	0.010**
Detection of target sentences	-0.298	1.807	0.083
Detection of distractor sentences	-0.265	1.289	0.210
Constant	0.000		

$N=32$; $R^2=0.641$; $F_{reg}=5.139^{**}$

We used a simple analysis of variance to test if children's classification according to degree of the recall's organisation yielded differences in the explicit and implicit main idea. Means and standard deviations in both dependent measures are separately displayed in Table 3. Results are slightly significant for explicit main idea ($F(2,31)=4.308$, $p<.05$), with no differences among the three groups according to the Scheffé test. Nevertheless, results are highly significant in case of implicit main idea (F(3,31) = 12.134, p < .001), and here the Scheffé test indicate significant differences in several cases (O > N, p < .05; p > N, p < .01).

Table 3: *Means and standard deviations of getting the explicit and implicit main idea for subjects classified according to their degree of recall's organisation (N = 32)*

	Explicit main idea		Implicit main idea	
	Mean	SD	Mean	SD
(N) Not organised (n = 22)	1.182	1.05	0.341	0.49
(P) Partially Organised (n = 5)	2.400	1.14	1.600	0.65
(O) Organised (n = 5)	2.200	0.44	1.200	0.83

The second hypothesis stated that both factors mentioned earlier, namely awareness of textual structure and ranking sentences, would also predict the recall of the most important information. To test this hypothesis we also carried out the two types of analysis mentioned above. First of all, Table 4 presents separately the means and standard deviations for recall of the three most and least important idea units for subjects classified according to the degree of the recall's organisation, the third task used to check textual structure awareness. The results of simple analysis of variance for the three most important idea units yields significant differences among groups (F(2,31) = 36.258, p < .001). *Post hoc* comparisons between them also showed significant differences in several cases (O > P and O > N, p < .01). In case of the low level idea units, significant but lower differences were found (F(2,31) = 7.667, p < .01), and differences were found only between children who organised their recall and those who did not organised it at all (p < .01).

We also used multiple regression in order to find the contribution of the variables mentioned earlier. These results are shown in Table 5. Only the variable 'matching sentences' is significant in case of the three most important idea units. Any of the variables were significant for the recall in the three least important idea units. This and the result mentioned earlier may support the importance of awareness of textual structure in recall of the most important information, at least with two of the measures, but it does not confirm this in case of the skill in ranking sentences.

Table 4: *Means and standard deviations of the three most and least important idea units recall for subjects, classified according to their degree of recall's organisation (N=32)*

	Most important		Least important	
	Mean	SD	Mean	SD
(N) Not organised (n=22)	0.318	0.56	1.019	1.60
(P) Partially Organised (n=5)	1.400	1.14	3.000	3.64
(O) Organised (n=5)	4.294	1.78	5.200	3.10

Earlier results show evidence about our third hypothesis which stated that reading microprocesses measured by some indicators of decoding would not predict reading macroprocesses. As far as getting the main idea from texts is concerned, the hypothesis is generally confirmed although the reading time variable is nearly significant for the second dependent variable (getting the implicit main idea) (see Table 2). In case of recalling the most important information from texts, it is fully confirmed by these data.

It should be remembered that our study had the additional purpose of comparing the results obtained with the different measures of textual structure awareness. So, apart from the differences among them that we have pointed out before, we also obtained the Pearson's and Spearman's rank correlation coefficients between each pair of measures (see Table 6). It should be noted that three of the measures (matching tests; detection of distractor sentences; and recall's organisation) are closely related although the figures are low if we consider that they are measuring the same skill.

Table 5: Results obtained in the multiple regression analysis for recall of the three most and three least important idea units from the text (* p<.05; ** p<.01)

Variables	Most important			Least important		
	Stand. coef.	T	p	Stand. coef.	T	p
Sums of errors in decoding	-0.172	0.622	0.53	-0.227	0.855	0.40
Sum of reading time	0.221	0.797	0.43	0.194	0.597	0.55
Verbal intelligence	0.600	1.692	0.10	0.407	0.980	0.33
Non-verbal intelligence	-0.082	0.282	0.78	0.010	0.028	0.97
Ranking sentences	0.169	0.974	0.34	-0.254	1.251	0.22
Matching texts	0.444	2.267	0.03*	0.130	0.567	0.57
Detection of target sentences	-0.055	0.297	0.76	0.231	1.071	0.29
Detection of distractor sentences	-0.158	0.689	0.49	-0.117	0.436	0.66
Constant	0.000			0.000		
N=32	R^2=0.555; F_{reg}=3.583**			R^2=0.389; F_{reg}=1.832		

Table 6: Correlation coefficients between measures of textual structure awareness (N=32) (a=Pearson's coefficients; b=Spearman's rank coefficients Rho corrected for ties; * p<.05; ** p<.01)

	Matching[a]	Target[a]	Distractors[a]	Recall's[b] Organis.
Matching texts	1.000	0.410**	0.584**	0.679**
Detection of target sentences		1.000	0.293	0.382*
Detection of distractor sentences			1.000	0.494**
Degree of recall's organisation				1.000

In fact, these correlation coefficients are very similar to what we found correlating 'verbal intelligence' or 'getting the implicit main idea' with 'matching texts'. 'Detection of target sentences' is the least related with the other measures, and even the correlation coefficient with 'detection of distractor sentences' is not significant.

Discussion

Results partially confirm our predictions. Our first hypothesis stated that awareness of textual structure and the skill in ranking ideas in a text would predict the ability of getting the main idea from expository text. Results show a clear-cut pattern depending on the fact whether the main idea was explicit or implicit. In the first case, the hypothesis may be disregarded because only one item of the textual structure awareness (degree of recall's organisation) produces slightly significant differences. This could be explained because of the small number of subjects classified as having partially and fully organised recall. It should be remembered that the text was quite long and that, above all, the content was unknown to children. Opposite to this, the hypothesis might confirmed because two measures of textual structure awareness (degree of recall's organization; matching texts) are strongly significant, and another one (detection of target sentences) is close to the usually admitted level of significance. Nevertheless, the other awareness measure (detection of distractor sentences) does not significantly predict any of the two dependent measures we are referring to. This will be commented upon later when we deal with the question of consistency among the different measures of awareness. The second part of our first hypothesis was related to the skill in ranking ideas. This measure does not predict the scores in getting the explicit main idea but it contributes significantly to predict the scores in getting the implicit main idea. This different pattern of results between texts with explicit and implicit main idea has also been found by Hare et al. (1989) in production tasks similar to our measures. They explain this fact in that the reader has to use an easier rule, the "selection" rule, in case of explicit main idea, but more difficult rules, such as "construction" or "generalisation" rules, in case of an implicit main idea. Therefore, mental processes seem to be different in both cases.

We have also found that verbal intelligence predicts significant results in getting the implicit main idea from texts. This was not expected but it is a current conclusion in studies about the relation between intelligence and reading comprehension (Harris & Sipay, 1980).

Our second hypothesis stated that two factors mentioned earlier (awareness of textual structure; ranking sentences) would also predict the recall of the most important information. Here, results obtained from different awareness tests are less coincident than before. Subjects who organised their recall according to the structure of the text recalled higher level information than those who did not organise their recall in this way. This results is coincident with other studies mentioned above (Meyer, Brandt & Bluth, 1980; Taylor & Samuels, 1984; Sanchez, 1988). Matching texts also contributes significantly to predict the scores of important information recall, but all other awareness measures do this. Probably, these other measures are more adequate to test comprehension than recall. In fact, detection of distractor sentences is similar to detection of inconsistencies, the most usual measure employed to test monitoring comprehension. Both task, comprehension and recall, are not completely independent, but they certainly imply different cognitive processes (Fisher & Mandl, 1984). It should be noted that results are different if we consider recall of the three least important idea units. In this case, 'matching text' does not predict the scores at all. Only the degree of recall's organisation explains the scores, but this could be due to the small number of subjects classified as having partially and fully organised recall, as we have pointed out above. Therefore, the role of textual structure awareness in recall is different depending on the level of importance which is congruent with the Van Dijk and Kintsch model.

The scores given to ranking sentences do not significantly contribute to recalling the most important information's scores. This is not the expected result. We think that this could be due to the difficulty of the text we employed. In fact, scores were very low. Therefore, the floor effect might explain these anomalous data.

Finally, our third hypothesis predicted that reading microprocesses measured by some indexes of decoding, such as reading time or errors in decoding, would not predict reading macroprocesses such as getting the main idea or recalling the most important information from texts. Obviously. it does not mean that micro- and macroprocesses are independent, but that the former ones are a necessary, though not sufficient, condition for the latter ones. This hypothesis has almost fully been confirmed. Only the sum of reading time reaches a score close to the 0.05 level of significance (0.076). Our results are coincident with current conclusions which find that correlation between decoding and comprehension decreases as school grades rise, being very low at high school (Harris & Sipay, 1980). This tendency could be accelerated in very phonetic languages like Spanish. In fact, Spanish fifth-graders normal readers have a decoding level very similar to adult readers (Cervera & Toro, 1979).

A second aim of our study was to compare different tests of textual structure awareness in expository texts. Most of the correlation coefficients are significant, but several questions should be pointed out. First, we have shown above that different awareness tests produce equally different results. Richgels *et al.* (1987) have arrived at a similar conclusion comparing "recall" and "comprehension" measures. Secondly, detection of target sentences is not significantly correlated with detection of distractors, and the correlation coefficients with the other two awareness measures are very low. Englert & Hiebert (1984) did not find differences between third- and sixth-grade school children with this measure, both groups scoring very high on it. Therefore, it seems that this measure is not very adequate for this metacognitive skill. Thirdly, the other three measures show high correlation indexes but they are very similar to correlation coefficients obtained between each one of the three measures and verbal intelligence or getting the implicit main idea from a text (range from 0.73 to 0.74). Therefore, awareness of textual structure seems to be a confusing construct.

In our opinion the problem is that expository prose cannot be constrained to have a definite class of textual structures. Therefore, as Kintsch (1988) claims in the last revision of his model, "prediction or expectation-based systems of comprehension that use frames or scripts do not adapt easily to new contexts" (p. 180). In fact, several studies have proved that adults and experts employ their knowledge of textual structures in a more flexible way than children or novices (Dee-Lucas & Larkin, 1986; Olhausen & Roller, 1988).

References

Anderson, T.H., & Armbruster, B.B. (1984). Content Area Textbooks. In R.C. Anderson, J. Osborn & R.J. Tierney (Eds.), *Learning to Read in American Schools*. Hillsdale, NJ: Lawrence Erlbaum Associates.

Armbruster, B.B., Anderson, J.H., & Ostertag, J. (1987). Does Text Structure/Summarization Instruction Facilitate Learning from Expository Text? *Reading Research Quarterly, 24*, 331-346.

Brooks, L.W., & Dansereau, D.F. (1983). Effects of Structural Schema Training and Text Organization on Expository Prose Processing. *Journal of Educational Psychology, 75*(6), 811-820.

Brown, A.L. (1980). Metacognitive Development and Reading. In R.J. Spiro, B.C. Bruce & W.F. Brewer (Eds.), *Theoretical Issues in Reading Comprehension*. Hillsdale, NJ: Lawrence Erlbaum Associates.

Cervera, M., & Toro, J. (1979). *Test de Análisis de Lecto Escritura (T.A.L.E.)*. Madrid: Visor.

Cook, L.K., & Mayer, R.E. (1988). Teaching Readers about Structure of Scientific Text. *Journal of Educational Psychology, 80*, 448-456.

Dee-Lucas, D., & Larkin, J.H. (1986). Novice Strategies for Processing Scientific Texts. *Discourse Processes, 9*, 329-354.

Englert, C.S., & Hiebert, E.H. (1984). Children's Developing Awareness of Text Structures in Expository Materials. *Journal of Educational Psychology, 76*(1), 65-74.

Englert, C.S., Stewart, S.R., & Hiebert, E.H. (1988). Young Writer's Use of Text Structure in Expository Text Generation. *Journal of Educational Psychology, 80*(2), 143-151.

Englert, C.S., & Thomas, C.Ch. (1987). Sensitivity to Text Structure in Reading and Writing: A Comparison between Learning and Disabled and Non-learning Disabled Students. *Learning Disability Quarterly, Spring*, 93-105.

Fisher, P.M., & Mandl, H. (1984). Learner, Text Variables and Control of Text Comprehension and Recall. In H. Mandl, N.L. Stein & T. Trabasso (Eds.), *Learning and Comprehension of Text*. Hillsdale, NJ: Lawrence Erlbaum Associates.

Hare, V.Ch., Rabinowitz, M., & Schnieble, K.M. (1989). Text Effects on Main Idea Comprehension. *Reading Research Quarterly, 24*(1), 72-88.

Harris, A.J., & Sipay, E.R. (1980). *How to Increase Reading Ability*. New York: Longman.

Kintsch, W. (1988). The Role of Knowledge in Discourse Comprehension: A Construction- Integration Model. *Psychological Review, 95*(2), 163-182.

Mayer, R.E. (1985). Structural Analysis of Science Prose: Can We Increase Problem Solving Performance? In B.K. Briton & J.B. Black (Eds.), *Understanding Expository Text*. Hillsdale, NJ: Lawrence Erlbaum Associates.

Meyer, B.J.F. (1984). Text Dimension and Cognitive Processing. In H. Mandl, N.L. Stein & T. Trabasso (Eds.), *Learning and Comprehension of Text*. Hillsdale, NJ: Lawrence Erlbaum Associates.

Meyer, B.J.F., Brandt, D.M., & Bluth, G.J. (1980). Use of Top Level Structure in Text: Key for Reading Comprehension of Ninth-grade students. *Reading Research Quarterly, 16*(1), 72-102.

Meyer, B.J.F. & Freedle, R.O. (1984). Effects of Discourse Type on Recall. *American Educational Research Journal, 21*(1), 121-143.

Olhausen, M.M., & Roller, C.M. (1988). The Operation of Text Structure and Content Schemata in Isolation and in Interaction. *Reading Research Quarterly, 23*(1), 70-88.

Richgels, D.J., McGee, L.M., Lomax, R.G., & Sheard, C. (1987). Awareness of Four Text Structures: Effects in Recall of Expository Text. *Reading Research Quarterly, 22*(2), 177-196.

Sánchez, E. (1988). Aprender a leer y leer para aprender: Características del escolar con pobre capacidad de comprensión. *Infancia y Aprendizaje, 44*, 35-57.

Scardamalia, M., & Paris, P. (1985). The function of Explicit Discourse Knowledge in the Development of Text Representations and Composing Strategies. *Cognition and Instruction, 2*(1), 1-39.

Taylor, B.M., & Samuels, S.J. (1983). Children's Use of Text Structure in the Recall of Expository Material. *American Educational Research Journal, 40*, 517-528.

Van Dijk, T.A., & Kintsch, W. (1983). *Strategies of Discourse Comprehension*. New York: Academic Press.

Vidal-Abarca, E. (1990). Un programa para la enseñanza de la comprensión de ideas principales de textos expositivos. *Infancia y Aprendizaje, 49*, 53-71.

Winograd, P.N. (1984). Strategic Difficulties in Summarizing Texts. *Reading Research Quarterly, 19*(4), 404-425.

4

Hypermedia and knowledge acquisition from non-linear expository text

Ala Samarapungavan
University of Amsterdam (The Netherlands)
Jos Beishuizen
University of Leiden (The Netherlands)

Abstract

This paper provides an overview of a current project that examines knowledge acquisition from expository text in complex domains. The main focus of this research is on how different text structures, implemented in a hypermedia environment, interact with the reader's prior knowledge to influence learning outcomes. In this paper, we will address some important theoretical issues in processing linear and non-linear expository text for learning in complex domains. We will then present some interim results from our ongoing research and discuss their implications for future research.

Introduction. Text structure and learning: Theoretical issues

Studying from expository text constitutes an important form of learning both within and outside formal education. In our culture, older pupils in particular are often required to learn from long expository texts such as "introductory texts" for college courses. Such texts serve as a source from which readers may learn about complex conceptual domains such as medicine and law. Over two decades of research on text processing have yielded much information about the cognitive processes involved in reading text, comprehending it, and remembering it. Research has shown that both prior knowledge of the conceptual domain circumscribed in the text (Van Dijk & Kintsch, 1983) and knowledge about the rhetorical structure and devices used by different genres of text plays a role in text comprehension (Meyer & Rice, 1984; Schnotz, 1984; Waller & Whalley, 1989).

Much of the research on expository text comprehension to date deals with short texts that contain approximately 500 words (Just & Carpenter, 1987). Furthermore, the expository texts studied often present a simple conceptual content of the kind encountered by readers in early stages of knowledge acquisition. Our own research program in contrast focuses on how readers learn from relatively long expository texts (20,000 words). We concentrate on texts that present information in complex knowledge domains.

A wide variety of text structures are currently in use for the presentation of large amounts of expository material ranging from loose encyclopaedic formats that do not suggest any particular ordering of text units in reading, to strict hierarchical formats that prescribe a definite reading sequence for text units. These text structures have been employed for both printed and electronically presented expository text.

Within the context of our research, we have chosen to focus on electronic reading environments for several reasons. First, reading text on computer is becoming an increasingly important activity in a variety of formal and informal learning contexts such as distance education, individualised learning environments, and information sharing and dissemination devices such as data bases and bulletin boards for business and scientific use. Secondly, certain technological developments such as the use of "hypertext" and "hypermedia" software have lead to the development of novel text formats that are not easy to accomplish with printed text. Finally, using electronic text presents certain methodological advantages by allowing us to automatically record certain theoretically important variables in the reading process such as which units of text a reader has explored, the time spent on reading each unit, and the sequence in which text units are read. Before we embark upon a more detailed discussion of the theoretical and empirical issues involved, we will provide a brief description of hypertext environments.

Hypertext

Hypertext is a generic name for a class of electronic reading environments. The special feature of such environments is that their software capabilities allow for various manipulations of text structure such as changes in the sequence of presentation of units within a text. These manipulations of text structure can be controlled either by the end user of the text or the designer of the text or both. For example, different readers can order units of the same text in different reading sequences because they can move from any unit in the text to any other directly at the click of a button. Readers may also be able to juxtapose two separate units of text on the screen in adjacent windows.

Some researchers have proposed that the novel electronic text formats are particularly effective in facilitating the acquisition of complex domain knowledge from large amounts of expository material. The term "hypertext" itself was coined by Nelson (1965) to denote a new "non-linear" text format that did not prescribe a specific reading sequence (and by implication a sequence for connecting ideas) through the information in a text. Implicit in this "democratic" model of interaction between the reader and the text is the notion that readers are capable of creating a coherent virtual text structure (or multiple structures) by selective browsing through units in the electronic text base. In this vision, long expository text is to be structured as a well- catalogued library of information that the reader could selectively explore. The most direct and currently most widespread application of hypertext based on this approach is the development of text bases that support information retrieval activities in experts. Such "texts" often perform a function similar to that of a printed reference manual. In contrast more general issues of learning or knowledge acquisition have been neglected.

Why non-linear text?

Little systematic research has been done that makes direct comparisons between the effects of reading from hypertext formats as compared to more conventional text formats for comprehension and learning of a target expository domain. Thus, there is scant evidence that the novel electronic text formats are more effective "cognitive tools" than conventional ones in fostering good reading and learning processes. Even if experiments were to show that systematic advantages increase for those who learn from expository text in hypertext formats we would still need to develop and test a theoretical model that could explain why and under what range of conditions such text structures could support effective knowledge acquisition. We propose that any such theory would have to consider the interactive roles of four major classes of variables:

(1) Domain variables which include decisions about what kinds of conceptual content or "knowledge" are to be presented and analyses of domain complexity and structure.

(2) Text variables which refer to both the content and the structure of the text itself as distinguished from the underlying structure of the domain being presented in the text.

(3) Learner variables which include prior domain knowledge, metacognitive skills related to reading and learning, and possibly individual difference variables such as preferences for holistic or serialistic modes of information processing.

(4) Task variables such as the goals that readers set for themselves or that are set for the reader through instruction.

In our discussion we would like to focus on aspects of two theoretical approaches which have served as a basis for recent research on the efficacy of hypertext structures in enabling readers to learn from expository text. Pask (1976) has suggested that a complex expository domain can be represented as an "entailment mesh" – a graphic device depicting concepts and conceptual relations in terms of a network of nodes and links.

One of the most common applications of Pask's theory in hypertext research has been as a rationale for the use of graphic menus or global network representations that summarise the conceptual content of the text for the reader in terms of a network of nodes and links. Within hypertext, such graphic devices become, along with direct functional connections or "jumps" between units, an active means of structuring text information, rather like the typographic and rhetorical devices used in conventional printed text. Much research shows that hypertexts which do use such global network representations of text content facilitate more systematic and coherent reading of the expository material than hypertexts which do not (Fosse, 1987). However, it is questionable whether such graphic devices can convey as much information about different kinds of complex conceptual relations to readers as the linguistic/rhetorical devices employed by more conventional expository text structures .

A second theoretical approach may be distinguished in the work of Spiro *et al.* (1987). They propose that knowledge in complex domains such as medicine, law, and history, is ill-structured and cannot be captured in a set of neat hierarchical relations. Concepts in complex domains are richly inter-connected in loops of reciprocal causation and multiple correlation. Additionally, concepts in complex domains are characterised by a high degree of exemplar variability so that choosing a single "classic" case to illustrate a concept is misleading. Finally, phenomena in complex domains typically need to be viewed from multiple perspectives for proper understanding. Drawing upon Wittgensteinian philosophy, Spiro *et al.* suggest that expository text in complex domains is best viewed as a landscape. In order to "know" the landscape one must take alternate paths through it and explore it through different perspectives.

There is research showing that even readers who are exposed to relatively well-structured and coherent expository (printed) texts show significant gaps in their understanding and mastery of the domain (Feltovich, Spiro & Coulson, in press). Spiro *et al.* (1987) have proposed that these problems stem partly from the limitations of conventional linear texts in conveying the nature of knowledge in structurally complex domains. This is because standard "linear" expository texts misrepresent knowledge in complex domains. Such problems may occur for two reasons: (1) Linear texts suggest a sequential flow of ideas through their typographic and rhetorical structure. As a result of this sequential presentation, topics that are conceptually related are often temporally and spatially separated for the reader. (2) Printed texts are limited by space constraints to the presentation of a few "text book" examples. An example once given is rarely revisited in other contexts. This hinders the consideration of domain phenomena from multiple perspectives.

Spiro & Jengh (in press) propose that non-linear texts of the kind implemented in hypertext environments can be used to improve the ability of readers to understand the conceptual organisation of complex domains. They claim that such non-linear texts reduce the likelihood of reductive errors made in linear text with regard to complex domains. Additionally, they suggest that non-linear text formats are also likely to be effective because they engage readers in an active "effort after meaning". For example, decisions about which text unit to explore next invite the reader to consider possible conceptual relations between text units and the juxtaposition of the same text unit with several other units can facilitate the construction of multiple perspectives on text information.

The theoretical perspective of Spiro and his associates is influenced by their interest in advanced knowledge acquisition. For example, Feltovich, Spiro & Coulson (in press) examine doctors' understanding of the bio-mechanical processes underlying cardiac arrest. Learners who are at such advanced stages of knowledge acquisition are perhaps more likely to engage and succeed in an active search for meaning in unstructured or loosely structured text. In contrast, readers with low domain knowledge or poor metacognitive skills related to reading and learning may not be able to comprehend or learn from highly unstructured or non-linear expository text. Spiro *et al.* (1987) found that low-ability high-school students found it hard to acquire coherent and structured knowledge form a non-hierarchical, case-based text on twentieth-century history. High-ability students, on the other hand, learned successfully from the non-hierarchical text.

Although we take the theoretical perspective of Spiro *et al.* as a point of departure for our own investigations, we believe some of the stronger claims for the efficacy of "non-linear" text in fostering the acquisition of complex knowledge and its future transfer to appropriate contexts should be viewed with caution. For one thing, we would like to make a theoretical distinction between the dimensions of "linearity" and "hierarchy" as the two are often used interchangeably in the existing literature on hypertext. Linearity, we suggest, should refer to the sequence of reading for text units suggested by a given text structure. For example, a list-like table of contents with topics and sub-topics, and linguistic connective devices within the text itself such as "In the following section ..." might strongly prescribe the order in which a reader ought to read text units. Hierarchy, on the other hand, should refer to the ordering of conceptual relations expressed within a text, for example, whether one concept has a causal or a part-whole relation to another concept.

The linearity of expository text may be partly determined by the hierarchical relations between domain concepts but this is not always the case. The reading sequence suggested for units within the text is not the only and possibly not even the most salient device for presenting hierarchical relations between domain concepts. The linguistic labelling of specific classes of conceptual relations such as counter-argument (e.g., "On the other hand ...") plays a very important role in cuing the reader to the hierarchical ordering of relations between domain concepts. We propose that neither linearity nor hierarchy should be viewed as dichotomies with regard to text structure. So called "linear" printed text is rarely strictly linear because rhetoric devices refer the reader to spatially distant but conceptually related text units, and readers are free to cycle back and forth between the units to explore such relationships. Conversely, even classic hypertext is not strictly non-linear because the reader never has access to all the material simultaneously and must impose some sequence of reading on the available units of text. Similarly, although some content domains may be characterised as very well-structured with a strict hierarchical ordering of conceptual relations and some as ill-structured, most domains fall in between.

Regardless of the specific devices through which expository text attempts to convey the structure of domain knowledge to readers, it will succeed or fail depending on its ability to represent the kinds of relations that adhere between domain concepts. If "non-linear" text or hypertext cannot provide the reader with clear and accurate information about the kinds of conceptual relations that characterise the target domain, it is unlikely to foster successful knowledge acquisition. As noted earlier, the outcomes of readers' interactions with text depend not only on the content and structure of the text itself but also on what readers bring with them in terms of their prior knowledge and their metacognitive activities during reading.

Our own research aims at contributing towards the development of a theoretical model of learning from expository text that accounts for variables of the kind we have discussed above. In the following section, we will describe some of the empirical work we have undertaken to examine the effects of linear and non-linear expository text on comprehension and learning.

Studying the effects of text structure in hypertext

Domain

The candidate domain we have chosen is the text on "memory" from Gleitman's (1986) *Psychology*, a widely used introductory text in cognitive psychology. A survey of 150 college professors (Weiten, 1988) showed that this was one of the five most highly rated textbooks in terms of pedagogical quality from a group of 43 texts examined. It was also the most highly rated in terms of level of discourse and quality of scholarship. We felt that in order to assess the effects of different text structures on comprehension and learning it was essential to select high quality texts for each type of structure. A further consideration was that Gleitman's *Psychology* was used as a textbook in the institution where the studies were conducted.

The first phase of research was concerned with the development of an expert conceptual model for the topic of memory through content analysis of the text on memory. Based on this analysis we designed manipulations of text structure for use in subsequent experiments.

Methods

The conceptual model
As mentioned above, the first step in our research was the domain analysis. Our goal was to develop an expert conceptual model identifying what a reader should know about the target domain based on our domain expertise and content analysis of the introductory text. The conceptual model identifies three types of elements described below.

The first type of element is the various perspectives from which topics can be viewed in the domain of memory. Four perspectives were identified as being represented in the text. These are:
(1) Phenomena – the kinds of events or behaviour that theories of memory typically address. A topic within this perspective would be the experience of the "tip of the tongue" phenomenon.
(2) Theories – this perspective refers to the broad theoretical approaches that have been applied not only to the domain of memory but to other domains of behaviour as well. A topic within this perspective might be "Associationism".
(3) Models – this perspective contains information with regard to specific theoretical models that have been proposed to handle phenomena in the domain of memory. A description of the three-stage information processing model would belong to this perspective.
(4) Experiments – this perspective presents information about the kinds of experiments conducted to study different aspects of memory. For example, the Ebbinghaus nonsense syllable experiments would belong to this perspective.

The second type of element in the conceptual analysis comprised a list of key topics and sub-topics that belonged to each of the four perspectives above. The third type of element identified in the analysis was the kinds of relationships that held between concepts in the domain (e.g., component to model, model to theory, experiment to model, etc.).

62

On the basis of the conceptual model we created an encyclopaedic structure for the information contained in the memory text. The units of text in this condition were accessed from a graphically organised menu called a "Map" of the domain (see Figure 1).

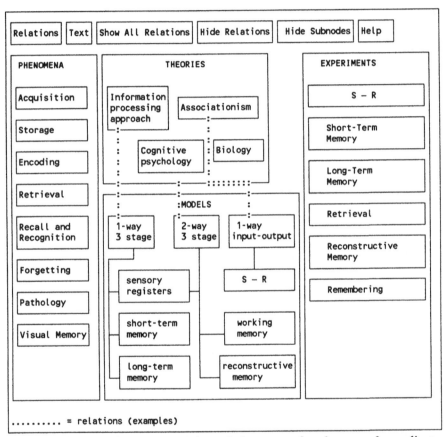

Figure 1: Schematic representation of the menu for the encyclopaedic text condition (Map)

The graphically organised menu represented the structure of conceptual relations determined in the conceptual model. Thus, the menu screen displayed four labelled and colour-coded background areas, each corresponding to one of the perspectives (phenomena, theory, model, and experiment) described above. Embedded in each perspective was a list of topics belonging to that perspective. If a reader chose to access one of these topics, he or she would then also have access to any sub-topics that might exist for that particular topic. The prose contained in the encyclopaedic text accessed through the "Map" was identical to that used in the original printed text with the exception of a few sentences that referred to the original sequential ordering of text units. The encyclopaedic structure was conceived of as being largely non-linear in the sense that there were no direct linguistic connections between individual units of text (accessed by menu entries) and also no graphic connections that prescribed a unique sequence for the reading of text units. However, the graphic organisation of the Map was designed to highlight some hierarchical relations between domain concepts (such as component to model and model to theory) by embedding topics in perspectives and indicating sub-topics for certain topics.

It was hypothesised that providing such information would constrain the selection of reading sequences. Thus, if subjects utilised the information provided in the graphic menu about the ordering and relation of concepts, they could still choose from among several useful reading sequences for text units but should not select other sequences possible in the reading environment.

After creating the non-linear "encyclopaedic" text structure we created a linear text structure. In this condition readers had access to an electronic version of the original linear text with the aid of the original "Table of Contents" as a menu to guide them in their selection of topics for reading (see Figure 2). We called this the linear condition because both the organisation of the menu entries and the rhetoric style of the text itself strongly suggested a given, linear reading path or sequence through the text.

Reader navigation through linear and non-linear text. In this phase we created a reader-controlled, electronic reading environment to study how differences in reader-selected paths through text influence knowledge acquisition in readers with different degrees of prior knowledge about the target domain. Although the experiments planned for this

64

phase are still in progress, we can report the results of a small pilot experiment.

Subjects

The subjects were 18 first-year undergraduate students in psychology who were novices with regard to the domain of memory. The subjects were randomly assigned to one of two text conditions (i.e., linear and encyclopaedic) to be described below. There were nine subjects in each condition.

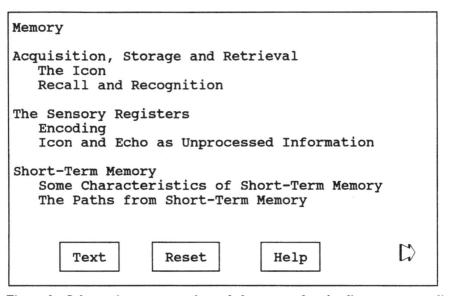

Figure 2: Schematic representation of the menu for the linear text condition (Table of Contents)

Text conditions

(1) Encyclopaedic text – In this condition subjects accessed the textual information through a graphic menu or "map" as described above. The graphically organised menu represented the structure of conceptual relations determined in the domain analysis (see Figure 1). The map provided information not only about what topics were related but about

the kinds of relations that existed (e.g., experiment to theoretical perspective).

(2) Linear text – The original text structure implemented electronically.

Procedure

Subjects were first shown how to use the electronic text. They were instructed to read the text in order to understand the domain of memory. They were told that at the end of the experiment they would be given a short multiple-choice test to determine their comprehension of the text. After the instructions had been given, all subjects read the complete text once, over a time period of three hours. The reading time was divided equally between two reading sessions scheduled one day apart. In both text conditions, the menu highlighted topics already read by the reader. The system kept a log of each subject's reading choices, providing information about the sequence in which information was sought and the time spent on each segment of information.

Data analyses

The performance measure recorded was the score on a short multiple-choice test of factual recall and inferential reasoning to measure reading comprehension. The subjects' scores on the DAT verbal analogies test were used to control for the effect of aptitude on performance. An analysis of covariance was performed with the subject's score on the reading comprehension test as the dependent variable, the text condition to which the subject was assigned as the independent variable, and the subject's DAT (verbal analogies) score as covariate.

Results

Because of the very small number of subjects it was not possible to draw any conclusions about the choice of reading paths or the traversal of text units within each text condition.

Reading comprehension test

As mentioned above, the subjects were also a given a series of multiple-choice questions to test their comprehension of the text. The questions were of two types: the first type of question was designed to test the subject's knowledge of important facts in the domain. For example, the subjects were required to identify what type of memory loss characterised the clinical condition of retrograde amnesia. The second type of question was designed to test the subject's ability to reason inferentially based on the information he or she had just read. For example, subjects would be given a short sentence and would be asked which words in it were likely to be remembered in chunks. Analyses of covariance were performed to determine whether there were any overall differences in performance on the reading comprehension test between the two text conditions. Further analyses examined whether such differences were specific to one of the two text components (factual and inferential). In the analyses, the subject's scores on the DAT verbal analogies test were entered as covariates to control for differences in performance due to the aptitude. The analyses showed that the covariate did not affect the dependent measures significantly. However, there was a significant difference ($p < .05$) in the overall performance of the novices between the two text conditions. The novices in the encyclopaedic condition performed better on the reading comprehension test as a whole than those in the linear condition. Further analysis showed that this difference was due to a difference in performance on the inferential reasoning questions. Novices in the encyclopaedic condition scored higher than those in the linear condition (see Table 1). There was no difference between the two groups on the factual items.

Table 1: Mean scores of novices on the reading comprehension test

	Text Condition	
	Linear	Encyclopaedic
Total score	12.17	17.00
Factual items	7.50	9.50
Inferential items	4.33	7.00

Discussion

Although the results presented above show the potential of certain non-linear text formats for facilitating knowledge acquisition from expository text, we believe that much further research is needed before we have a complete theoretical model of how alternate text formats affect knowledge acquisition among different groups of readers. It should be noted that the novices in our research were all university students who did not differ significantly on general ability measures (such as performance on the DAT verbal analogies test). They all evidenced good recall of factual information in text regardless of the text condition to which they were assigned. It is likely that the population of university students comes equipped with fairly advanced reading and study skills that allow them to learn successfully in novel and demanding conditions such as our encyclopaedic text conditions. However, the effort after meaning, that non-linear text structures (even well-structured ones such as ours) demand may place an excessive cognitive load on low-ability readers and learners. All our subjects in the encyclopaedic condition mentioned that they found their reading condition to be very difficult and tiring.

A second aspect to consider is whether the domain itself is suitable for presentation in a non-linear expository format. Related to this general issue are questions of the degree to which the domain is conceptually complex and/or ill-structured, and the degree to which it may necessitate some simplification or reduction for novice readers. Although our domain lent itself quite well to a well-structured, non-linear text format this may not be possible for every expository domain.

In our current research we have only considered what people learn from a single reading of long expository text. However, in real-life contexts, learning from such expository texts involves multiple readings of the same text. In future experiments we plan to examine how multiple readings of linear or encyclopaedic texts, or combinations of both, affect learning outcomes.

Our research has been primarily concerned with a relatively broad definition of knowledge acquisition from expository text. We have been examining how people learn when they read a text in order to generally "understand" or know about a domain. Although this is also the most common pedagogical goal for expository text in educational contexts, readers' more specific goals (such as studying for a multiple-choice exam) are also likely to influence how they learn from different kinds of text formats.

Finally, it is possible that individual difference variables such as Pask's (1982) notion of preferences for serialist versus holist search for information might also play a role in how readers acquire knowledge from different types of expository text (Van der Veer and Beishuizen, 1988).

Conclusions

The current study replicated our earlier findings with a similar group of novice subjects that well-structured encyclopaedic text formats which provide opportunities for conceptually constrained non-linear reading can enhance a reader's understanding and knowledge of the expository domain. Of particular interest is the finding that while there is no difference in the recall of facts across good exemplars of linear and non-linear text formats of the kind we have employed in our research, there are differences in the ability to make inferences based on the text information in novel contexts.

References

Feltovich, P.J., Spiro, R.J., & Coulson, R.L. (in press). The nature of conceptual understanding in biomedicine: The deep structure of complex ideas and the development of misconceptions. In D. Evans and V. Patel (Eds.), *The Cognitive Sciences in Medicine*. Cambridge, MA: MIT Press.

Fosse, C.L. (1987). *Effective browsing in hypertext systems*. CeRCLe Technical Report No. 41. University of Lancaster.

Gleitman, H. (1986). *Psychology*. New York: Norton.

Just, M.A., & Carpenter, P.A. (1987). *The psychology of reading and language comprehension*. Newton, MA: Allyn & Bacon.

Meyer, B.J.F., & Rice, G.E. (1984). The structure of text. In P.D. Pearson (Ed.), *Handbook of Reading Research*. New York: Longman.

Nelson., T.H., (1965). *A file structure for the complex, the changing, and the indeterminate*. Proceedings of the ACM National Conference, pp. 84-100.

Schnotz, W. (1984). Comparative instructional text organization. In H. Mandl, N.L. Stein & T. Trabasso (Eds.), *Learning and comprehension of text*. Hillsdale, NJ: Lawrence Erlbaum Associates.

Spiro R.J., & Jengh, J.C. (in press). Random access instruction: Theory and technology for the non-linear and multidimensional traversal of complex subject matter. In D. Nixand & R.J. Spiro (Eds.), *"The Handy Project": New Directions in Multimedia Instruction*. Hillsdale, NJ: Lawrence Erlbaum Associates.

Spiro, R.J., Vispoel, W.P., Schmitz, J.G., Samarapungavan, A., & Boerger, A.E. (1987). Knowledge acquisition for application: Cognitive flexibility and transfer in complex content domains. In S.M. Glynn & B.C. Britton (Eds.), *Executive Control Processes in Reading*. Hillsdale, NJ: Lawrence Erlbaum Associates.

Van Dijk, T., & Kintsch, W. (1983). *Strategies for Discourse Comprehension*. New York: Academic Press.

Veer, G.C. van der, & Beishuizen, J.J. (1988). Computers and education: Adaptation to individual differences. In G.C. van der Veer, T.R.G. Green, J.-M. Hoc & D.M. Murray (Eds.), *Working with Computers: Theory Versus Outcome*. London: Academic Press.

Whaller, R., & Whalley, P. (1989). Graphically organized prose. In E. De Corte, H. Lodewijks, R. Parmentier, & P. Span (Eds.), *Learning and Instruction*. Leuven University Press.

Weiten, W. (1988). Objective features of introductory psychology textbooks as related to professors' impressions. *Teaching of Psychology, 15*(1).

5

Text comprehension and typographical features of elementary school textbooks[1]

Pietro Boscolo, Lerida Cisotto and Albina Lucca

University of Padova (Italy)

Abstract

Research on text design has flourished in the last decade, mainly focusing on the function of adjunct questions, summaries, pictures, and so on, while, until recently, little attention had been paid to typographical organisation as a relevant variable in instructional text processing. The aim of this study was to investigate the effects on learning from text of a typographical device which is extensively used in elementary school textbooks: the box, i.e., a frame in which some information is highlighted. 140 fourth-graders read an expository text on insects, in which the positioning of the box and the type of framed information varied. After reading, they were submitted to a multiple-choice test and a written recall task. Different MANOVA models were adopted and the results showed that the presence and positioning of the box had no influence either on literal or comprehension item responses. On the basis of log-linear analyses it emerged that the effectiveness of the box seems to depend both on the type of information framed in it and on the learner's comprehension ability level.

---

[1] The preparation of this paper was supported by a grant from the National Council of Research (CT 89.02071.08).

Introduction

The design of an instructional text is a complex enterprise which concerns not only text organisation and structure, but also how written discourse can be most effectively presented to the learner. By effective presentation we mean the design and layout of the printed page, i.e., the technology of text (Jonassen, 1982; 1985; Hartley, 1985; Gropper, 1991). The technology of text can be defined as the application of a scientific approach to text design, as "a counterpoint to the artistic and unsystematic approach" to traditional text design (Jonassen 1982, p. x). Research on text technology has flourished in the last decade and has focused mainly on the function of devices such as adjunct questions, summaries, illustrations, and their effectiveness in facilitating learning. In the sixties E.Z. Rothkopf formulated the mathemagenic hypothesis, according to which certain design attributes of any medium give rise to learning: the form or structure of the display induces the cognitive processes which underlie learning (see Rothkopf, 1970, 1982, for a synthesis of the hypothesis). In other words, mathemagenic aids cue or encourage learners to cognitively operate on information in a way that they would not under "natural" conditions. So, according to this hypothesis, which has been mainly supported by Rothkopf's extensive research on adjunct questions, the learner can be oriented to the text, or some part of it, by cues which allow a control of the learning process. The mathemagenic hypothesis has only been partially confirmed by successive research, while in the seventies a cognitive view of learning emerged, according to which a learner, when faced with instructional stimuli, constructs and assigns meaning to that information on the basis of his/her prior learning. However, the cognitive approach to learning from text has focused essentially on text structure and task context variables, and has rather neglected aspects of text design. Until recently, for instance, little attention has been paid to the typographical organisation of text as a relevant variable of instructional text processing. Most studies on this topic are due to J. Hartley (e.g. 1980, 1985, 1986, 1987, etc.). Hartley has investigated several aspects which should be taken into account in planning the typographical structure of a text, such as page size, positioning of illustrative materials, margins and columns, headings, etc. A very common typographical device such as the heading, for instance, has two main instructional functions (Hartley and Jonassen, 1985). First, it facilitates the initial encoding of information from the text into memory; secondly, it facilitates access to the text for the reader who is either re-examining a familiar text or searching through a new one.

Few studies, if any, have been conducted on the effects on text comprehension and retention of a typographical device which is very common, especially in elementary school textbooks: the box. The box is a frame in which some information is highlighted. Although the box is very widely used in elementary textbooks, its effectiveness as an instructional aid still has to be demonstrated. Research is particularly lacking on the effects of framing different types of information. In fact, a box is typically used as a reminder, i.e., to highlight the most important ideas of a passage (e.g., the essential features of the object/phenomenon described). In this case the box should help the reader synthesise and therefore is usually situated at the end of the passage. In other cases the box presents some curious information, which is not particularly relevant but can make the passage more "interesting" or appealing. In these cases it can be considered a reminder as well, but of a different type. When highlighting relevant information it is intended to stimulate or facilitate the reader's synthesis of passage information. In the case of a curiosity, information is to be retained rather than elaborated so the box is usually situated in a more lateral position. Therefore, the function of a box seems to vary both in its position on the page and, particularly, in the type of information it frames. Consequently, a distinction between comprehension is made in the general expression "learning from text". By comprehension we mean the elaboration of text information by means of inferential processes based on the reader's knowledge, while retention regards information as stated in the text (Cunningham, 1987).

This study has been conducted according to two main hypotheses. The first hypothesis is a general one and refers to the concept of mathemagenic activity: framing some part of a text makes learning the global material easier than if the same text has no box. Moreover, a different effect is expected according to the learners' comprehension ability level.

The second hypothesis specifically concerns the effects of the box on the comprehension of the information it presents. We hypothesised a differential effect according to: (a) the central or marginal position of the box, which is usually located either in the middle or on one side of the printed page, thus giving a different relevance to the framed information; (b) the type of information inserted in the box, either for its relevance (for instance, the defining traits of a concept) or for its interest value, e.g., a curious anecdote.

Method

Measures and procedure

A 350-word passage was prepared in which the physical characteristics and main classifications of insects were described. Text information was organised as a description (Meyer, 1975), i.e., as a sequence of paragraphs, each presenting some information on insects. The layout of the passage was similar to a textbook page, with three colour pictures on the right side of the page presenting the anatomy of insects, phases of metamorphosis and a particular species of insects quoted in the text. While the global information and the pictures remained the same across all the experimental conditions, the printed page varied according to the following dimensions:

(1) the position of the box, which could be situated in the middle or on the right side of the page;

(2) the type of information in the box, which could be:

    (a) the main features of insects;

    (b) the meaning of "metamorphosis" (terminology);

    (c) some details of an insect living among the pages of old books (curious detail).

There was also a no-box condition, in which the passage was presented with no framing.

A multiple-choice test was prepared to assess the subjects' comprehension of the passage. It consisted of 15 items, of which 7 assessed the retention of verbatim information, and 8 assessed comprehension, i.e., required an inferential process and a paraphrase (Anderson, 1972). There is an example of the two types of items in Table 1.

The passage was read by the subjects during normal school time. The next day they were asked to write all that they could recall about the passage, and the following day they carried out the multiple-choice test. The passage was divided into information units (see content analysis).

Table 1: Examples of test items assessing subjects' comprehension of text passages

---

Information in the passage:
These animals have two compound eyes, i.e., formed by many small eyes. On the insects' head there are also two antennae which function as sense-organs (hearing, olfaction, etc.) and the mouth apparatus.

Example of verbatim or literal item:
Insects' eyes are formed by:
  • two antennae which serve to see
  • many small eyes
  • two small holes on the two sides of the body

Example of comprehension item:
Insects' antennae serve:
  • to suck nectar from flowers
  • to fight against other insects
  • to detect noise, smell, danger

Subjects

The subjects were fourth-graders from an elementary school in a town near Padova, who had never previously been exposed to any oral or written explanation of the insect topic. They were chosen on the basis of reading comprehension test scores (Cornoldi *et al.*, 1981) and divided into two groups: high and medium-level readers. Ten high-level and ten medium-level subjects were assigned to each cell of a 2 (box position) ×

3 (type of information) design. Moreover, the same number of subjects (10 + 10) was assigned to the no-box condition. There were 140 subjects in all.

Results

Multiple-choice test

First, the results of the multiple-choice test were analysed. Within the test two subtests were considered: the first included the items which required literal responses (7 items) and the second contained the remaining 8 items which required responses mainly based on the subjects' comprehension. The proportion of correct responses in the first and the second subtest were calculated separately.

A MANOVA model (Mardia, Kent & Bibby, 1979) was adopted in order to study the effect of the typographical device of the page (factor B with 2 levels: no-box and box) and the learners' comprehension ability level (factor A with two levels: high and medium) on the proportion of correct responses within both the subtests. There was a significant effect of A ($F = 77.12$, d.f. $= 2$, 107; $p < .0001$), but no significant effect of B.

Two ANOVAs,[2] one for literal responses and the other for comprehension responses, were calculated with the same factors. Only the ability level showed to be significant for both kinds of responses ($F = 59.54$, d.f. $= 1,108$; $p < .0001$ for the literal items; $F = 60.32$, $p < .0001$ for the comprehension items). Point estimation showed the highest number of correct responses among the high-level learners.

Secondly, another MANOVA model was adopted with three factors: ability level (factor A), position of the box (factor P with two levels: central and lateral), and type of information inserted in the box (factor I with three levels: i_1 – main features of insects, i_2 – terminol-

---

[2] The significance level for MANOVA was set equal to 5% and for each of the ANOVAs was set equal to 2.5%, following Bonferroni inequality (Miller, 1985; Vasey and Thayer, 1987; Huberty and Morris, 1989). Following Scheffé (1959), also overall results significant at 10% level were considered and reported as results tending to be significant.

ogy, i_3 – curious detail). Table 2 presents proportion means of correct responses for literal and comprehension items for the 2 ability levels crossed with the 2 different positions and the 3 types of information. Again, the ability level was highly significant, while the position of the box was not significant. The third factor I tended to be significant ($F = 1.95$, d.f. $= 4,214$; $p < .10$). Two ANOVAs were carried out for each of the two subtests. No significant result emerged regarding P, while the significant effect of A was confirmed. The most interesting result concerned factor I: the effect tended to be significant for literal responses only ($F = 3.38$, d.f. $= 2\ 108$; $p < .05$). The point estimation showed the highest number of correct literal responses for i_3 and the lowest for i_1.

Table 2: Proportion means of literal (L) and comprehension (C) item responses

		Central Box			Lateral Box		
		i_1	i_2	i_3	i_1	i_2	i_3
High	L	0.82	0.92	0.86	0.88	0.92	0.90
ability	C	0.92	0.92	0.84	0.94	0.91	0.87
Medium	L	0.66	0.62	0.75	0.65	0.71	0.78
ability	C	0.68	0.67	0.69	0.73	0.72	0.67

Thirdly, within the multiple-choice test three subsets of items were considered concerning the target information, i.e. the type of information included in the box. The first subset (C_1) included two literal and two comprehension items regarding the defining traits of the concept of insect, the second (C_2) included two items regarding the meaning of "metamorphosis", the third (C_3) included only one item regarding the curious detail of the insect topic.

A two-way log-linear model (Andersen, 1979) was adopted in order to study the interaction of B (box, no box) and the category responses for each subset for the two ability levels of learners.

For subset C_1 the response categories were defined as follows: r_{11}: bad responses (no correct responses or only one correct response); r_{12}:

medium responses (2 correct responses); r_{13}: good responses (3 and 4 correct responses). No significant effect emerged.

The same log-linear model was adopted for subset C_2, for which the following response categories were considered: r_{21}: bad responses (no or one correct response); r_{22}: good responses (2 correct responses). Again, no significant result was shown.

Only two response categories were defined in subset C_3: r_{31}: no correct response; r_{32}: correct response. No significant result emerged.

A three-way log-linear model was used to study the effects of the first and second order interaction of P (central and marginal box) × I (type of information) × R (response categories to each of the three subsets). The response categories were the same as those previously defined. The log-linear analysis was conducted for each of the two samples of learners.

For the high ability level subjects, the only result which tended to be significant was the interaction of I x R_1 (response categories for subset C_1) (chi-square = 8.62, d.f. = 4; p < .10). Post-hoc analyses (Andersen, 1980; Cristante & Lucca, 1988) of the single interaction parameters showed an interaction estimate for i_1 by r_{13} (of negative sign) significantly different from zero (z = -2.03; p < .02). In other words, for the high-ability subjects good responses to items of subset C1 were less frequent than expected when information in the box regarded the defining traits.

For the medium-ability level the only significant effect (chi-square = 6.15, d.f. = 2; p < .05) concerned the interaction of I x R_2 (response categories for subset C_2). Post-hoc analyses showed a positive interaction estimate significantly different from zero for i_2 by r_2 (z = 2.4; p < .01). In other words, for subset C2 of the items regarding terminology, for the medium-ability subjects, good responses were more frequent when the box included this type of information. No significant effect emerged for subset C_3 for both the ability levels of learners.

Written production

Content analysis
The passage was divided into 31 information units, i.e. into items which conveyed information on the topic of insects. From the syntactical point of view, a unit corresponded to a whole sentence or part of it.

Statistical analysis
As has already been described, the subjects' written productions were scored by giving 1 point to each sentence corresponding to a unit of the passage, i.e. for each subject every unit of the passage was scored as either 1 or 0 according to its correct presence in his/her written production.

First, the influence of the box/no-box condition on the written productions was studied. As in the multiple-choice test, no significant result emerged for the two groups of learners neither for the score of the full passage nor for the score concerning the target information, i.e., the units of the passage included in the box.

Secondly, the type of information was considered. For this purpose, the units of the passage included in the box were divided according to information type into: I_1 (main features of insects), I_2 (terminology), and I_3 (curious detail). An ANOVA model with three factors: P (position of the box), A (ability level), and I (types of information) was adopted for the analysis of the full passage scores (see Table 3). Only the ability level of the learners showed a significant result ($F = 71.67$, d.f. $= 1,108$; $p < .0001$).

Table 3: Proportion means of the information units correctly recalled in the written productions

	Central Box			Lateral Box		
	I_1	I_2	I_3	I_1	I_2	I_3
High ability	0.54	0.66	0.53	0.59	0.72	0.60
Medium ability	0.38	0.34	0.32	0.37	0.39	0.42

The subjects' written production units were also distinguished into the three subsets C_1 (main features), C_2 (terminology), and C_3 (curious detail). A three-way log-linear analysis was performed on each subset of information units of written productions.

First, subset C_1 was considered. Three categories of the units concerning C_1 were defined: C_{11} (0, 1 and 2 units correctly present in the written productions), C_{12} (3 units), and C_{13} (4, 5 and 6 units). The interaction of the two box positions (P) with the 3 types of information I (I_1, I_2, and I_3) and the 3 categories of written productions (W) within C1 was studied for both the ability levels. As in the multiple-choice test an overall I x W significant interaction emerged for the high ability level only (chi-square = 12.04, d.f. = 4; p < .02), while the P x I x W interaction was not significant. The post-hoc analysis concerning the single interaction parameters showed a negative interaction estimate which was significantly different from 0 for C_{13} x I_1 (z = -2.33; p < .01) and a positive interaction estimate for C_{11} x I_1 (z = 3.04; p < .001), confirming the results found in the multiple-choice test.

Then the log-linear analysis was performed for C_2 and C_3, creating the following categories for both the subsets: C_{21} and C_{31} (0, 1 unit), C_{22} and C_{32} (2 units), C_{23} and C_{33} (3 units). No significant result emerged for either ability level.

Discussion

The first and more general hypothesis of this study (i.e., of an overall effect of typographical cuing on retention and comprehension of passage information) has not been substantially confirmed, although a tendency to significance of the type of information factor in the literal item responses emerged. This tendency, which should obviously be verified with other passages and longer multiple-choice tests, seems to suggest that highlighting the information which is more relevant, or central, in a passage, such as the defining traits of the main topic, could be less useful for the literal retention of the passage than the highlighting of less relevant information. In other words, the process of synthesising information concerning the defining features gains little or no advantage from being highlighted, while the less relevant information does.

The results of analysis of target items seem to confirm this trend in relation to the comprehension ability level of the subjects. In fact, while the high-ability-level subjects gain less advantage from highlighting relevant information, for the medium-level subjects the crucial type of information seemed to be terminology, in which they got the highest scores. So, while the position of the box has no influence either on the literal or comprehension responses, the effectiveness of the box seems to depend both on the type of information it contains and on the learners' comprehension ability levels. More specifically, the box seems to have the function of signalling rather than fostering a deep elaboration of text information. This might suggest the opportuneness of using the box as an aid to information retention rather than information synthesis.

This main result of the study, i.e., the different effects of the box depending on the type of information inserted in it, puts forward two questions for future research on this topic.

The first question concerns the influence on readers' text processing of several variables (e.g., the length of the framed text, the presence of signalling devices such as "do you know that?", etc.), which have been only partially controlled in the present study. To this end, we are carrying out a study in which the different types of boxed information are presented to the subjects as homogeneously as possible.

The second question is a wider one, and concerns the role of typographical cuing in text processing. While it seems obvious that some features of layout may influence text processing, a theory of the relations between visible text features and semantic information processing is still lacking. The results of this study seem to suggest that highlighting information affects text processing at a surface rather than a deep level. The distinction between "surface" and "deep" level regards the relevance of sub-topics referring to insects. No attempt was made in the present study to relate the framed-information variable to any theory of text structure. The passage used was organised as a description, according to Meyer's (1975) classification of rhetorical predicates. It would be interesting to analyse the effects of highlighting information not only according to its conceptual relevance but also its function in a different, more complex text organisation. If highlighting text information can by no means be considered a panacea for comprehension, the conditions which could make highlighting more effective deserve to be studied not only in order to improve text design, but also to get a more comprehensive view of learning-from-text processes.

References

Andersen, E.B. (1979). *Discrete Statistical Models with Social Science Applications.* Amsterdam: North-Holland.

Anderson, R.C. (1972). How to construct achievement test to assess comprehension. *Review of Educational Research, 42,* 147-170.

Cornoldi, C., Colpo, G., & Gruppo, M.T. (1981). *La verifica dell'apprendimento della lettura.* Firenze: O.S.

Cristante, F., & Lucca, A. (1988). Estimation and test procedures for association parameters based on log-linear models in a contingency table of two psychological variables with nominal and ordinal categories. *Contributi di Psicologia, 1,* 144-156.

Cunningham, J.W. (1987). Toward a pedagogy of inferential comprehension and creative response. In R.J. Tierney, P.L. Anders & J.N. Mitchell (Eds.), *Understanding Readers' Understanding: Theory and Practice* (pp. 229-253). Hillsdale, NJ: Lawrence Erlbaum Associates.

Gropper, G.L. (1991). *Text Displays.* Englewood Cliffs, NJ: Educational Technology Publications.

Hartley, J. (Ed.) (1980). *The Psychology of Written Communication.* London: Kogan Page.

Hartley, J. (1985). *Designing Instructional Text* (2nd edition). London: Kogan Page.

Hartley, J. (1986). Planning the typographic structure of instructional text. *Educational Psychologist, 21,* 315-332.

Hartley, J. (1987). Typography and executive control processes in reading. In B. Britton & S. Glynn (Eds.), *Executive Control Processes in Reading* (pp. 57-79). Hillsdale, NJ: Lawrence Erlbaum Associates.

Hartley, J., & Jonassen, D.H. (1985). The role of headings in printed and electronic text. In D.H. Jonassen (Ed.), *The Technology of Text* (vol. 2) (pp. 237-263). Englewood Cliffs, NJ: Educational Technology Publications.

Huberty, C.J., & Morris, J.D. (1989). Multivariate analysis versus multiple univariate analysis. *Psychological Bulletin, 105,* 302-308.

Jonassen, D.H. (Ed.) (1982-1985). *The Technology of Text* (Vols. 1-2). Englewood Cliffs, NJ: Educational Technology Publications.

Mardia, K.V., Kent, J.I., & Bibby, J.M. (1979). *Multivariate Analysis.* London: Academic Press.

Meyer, B.J.F. (1975). *The Organization of Prose and Its Effects on Memory.* Amsterdam: North Holland.

Miller, R.G. (1985). *Simultaneous Statistical Inference*. New York: Springer Verlag.

Rothkopf, E.Z. (1970). The concept of mathemagenic activities. *Review of Educational Research, 40*, 325-336.

Rothkopf, E.Z. (1982). Adjuncts aids and the control of mathemagenic activities during purposeful reading. In W. Otto & S. White (Eds.), *Reading Expository Material* (pp. 109-138). New York: Academic Press.

Scheffé, H. (1959). *The Analysis of Variance*. New York: Wiley.

Vasey, M.W., & Thayer, J.P. (1987). A multivariate solution. *Psychophysiology, 24*(4), 479-486.

6

A procedure for the design of illustrated texts

Egbert Woudstra and Cees Terlouw
University of Twente (The Netherlands)

Abstract

We have developed a procedure for writing texts with the focus on illustrations. Our theoretical framework was twofold: we used a text theory and a cognitive process theory of text design. In the text theory a functional approach was used, distinguishing between subfunctions as question-description subfunctions, answer-information subfunctions, channel subfunctions and introduction/summary subfunctions. For our cognitive process theory of text design we used the original Flower and Hayes model. From this model and from the text theory we derived a systematic procedure and tested it. The procedure was tested with students by means of interviews and questionnaires. In the first phase of our research the writing procedure was only partially satisfactory. Meta-awareness of the procedural knowledge was improved and as a consequence the students produced texts of good quality. But the step by step approach did not sufficiently support the students in their planning and writing. In the second phase of our research we modified our theoretical framework to overcome our problems. In the cognitive process theory we included problem solving, reflective and intuitive actions. We consequently improved our systematic procedure. We limited our research to the design of a text scheme as a first step to a complete written text. This approach supported the students designing a text scheme including illustrations better. The development of our theoretical framework and the procedures will be discussed.

Introduction

The use of illustrations in several genres of educational texts is crucial. The presence of illustrations in, for example, handbooks, encyclopaedia, dictionaries, teacher's guides, directions for use for students, descriptions of products and processes to be understood by a lay audience, has the function that someone will comprehend the message of the text more adequately.

In our opinion the use of illustrations as an integral part of texts is a neglected and difficult aspect in the designing/writing process of a text. Especially in the text genre of 'illustrated popular-science text', the integral use of an illustration in the text is a real problem.

From interviews with popular-science writers and editors of popular-science journals we learned that in most cases there is a strong division in tasks between writers and editors. Writers are very much text-oriented. They believe that scientific concepts should primarily be explained verbally. Editors are much more illustration-oriented and lay-out-oriented (Ikink, 1987).

We think that a text can be more effective if the (content) writer knows how and when to use illustrations as an integral part of texts. For that reason we developed a systematic procedure in order to support writers in their illustrated text designs. At the same time it could be useful in simplifying the communication between writers and editors.

Our ultimate research objectives are:
(1) to develop a valid designing/writing model for an illustrated (popular) science text;
(2) to develop a usable systematic procedure for designing and writing an illustrated (popular-)science text;
(3) to develop an instructional setting in which the designing/writing of an illustrated popular-science text with the help of the systematic procedure can be learned by students of every discipline in higher education.

In this paper we restrict ourselves to the second research objective. However, because of the research setting, we also gathered some information on the two other research objectives.

Our explorative research question is the following: How can our systematic procedure, used in a concrete instructional situation, be designed in accordance with the theoretical frameworks as specified, and how must this procedure be improved? We will focus in particular on the help of the systematic procedure for constructing a text scheme.

The use of a systematic procedure as an educational tool originated in theories concerning problem solving. Therefore, we will develop a theory of text design from the perspective of problem solving (Mettes & Gerritsma, 1986; Terlouw, 1987; Goel & Pirolli, 1989; Hamel, 1990). Flower & Hayes (e.g., 1980, 1981) proposed a writing model based on the problem-solving theory of Newell & Simon (1972).

However, the model has been criticised for its many limitations concerning the linguistic aspects, the problem-finding aspects, the knowledge structure and the prescriptive value (Vanmaele & Lowyck, 1990). Furthermore, Stotsky (1990) criticises the process/product dichotomy assumption of the model: writing is only a product of the more important thinking process, and the assumption that the components of the writing process correspond to Newell & Simon's components in the process of solving problems in logical, spatial and quantitative relationships. Both assumptions result in a conceptual ambiguity concerning the key term "writing plan".

The criticism can be divided into a linguistic and a psychological part. Therefore, our theoretical framework must be twofold: one concerning the text theory, a theory about content and structure, and one concerning a cognitive process theory of text design, a theory about the design process involved. However, in this article we will mostly focus on the text theory.

In the first phase of our research we stressed the linguistic aspect by developing a text theory as a basis for analysing and writing popular-science texts (study one), taking into account the criticism on the original (psychological) model of Flower & Hayes concerning the content aspect.

Starting from the text theory we constructed a concrete instructional aid in the form of a procedure (study one). We conducted a study within this framework in an educational setting: students from engineering departments had to work with our approach in a course on science journalism. After having analysed the results of this experiment concerning the integrated use of illustrations and knowing the criticism on the psychological model of Flower & Hayes, we adapted this model and the procedure. In the second phase of our research (study two) we conducted some new experiments in more or less the same educational setting.

In this article we will first present study one (section 2); our text theory will be described in the introduction of this section. In section 3, we will present study two, the systematic procedure we constructed. In each section information about subjects, materials, procedure and results will be given. Each section will be ended with a short discussion section. We conclude with a general discussion in section 4.

Study one

The Flower & Hayes writing model

As mentioned above, we used the original, well-known writing model of Flower & Hayes. The model consists of three interactive blocks. One is called the task environment, one the long-term memory and one the working memory. In the working memory there is the distinction between, for instance, generating and structuring of information, writing and revision. The whole process is regulated by a monitor.

Functional text theory

The criticism on Flower & Hayes about the linguistic aspect concerns the lack of specific attention for the content of writing and the characteristics of the text genre to be designed. Each text genre has its own contents, structures and linguistic conventions, which are described in a text theory. Our text theory which is relevant for illustrated texts and especially popular-science texts is the following:

We see a text as a hierarchical structure based on a text theme and constructed along the question-answer principle (Steehouder *et al.*, 1984; Anderson & Armbruster, 1985). Some question-answer formats can be subordinate to others. Considering a text as a conglomeration of question-answer formats gives us the opportunity to relate frequently used questions directly to the topic of the text (the text theme), in terms of an object, a process or a theory. The level of such a question-answer format in the text hierarchy can be important for the decision how vital certain information is in the text. If a restricted space for illustrations is available (which is almost always the case), text hierarchy can play an important role in the decision to use a question-answer format in the form of an illustration.

Because we emphasise the aspect of an integrated text, we chose the concept of information as a basic principle. The information in a question-answer format can be given either verbally, in the form of an illustration or in a combination. The information itself can be structured in (a combination of) the following ways: spatially, functionally, attributionally, temporally, causally and comparatively.

We have chosen a functional approach. We define function in the area of illustrated text designing/writing as the result of the planned activity of the writer – with the help of subfunctions – in representing topic information to attain a planned effect. For reasons we set for our type of text, the overall goal concerning comprehensibility and attractiveness is implied.

Following the question-answer format we basically have *question-description subfunctions* and *answer-information subfunctions*. In the actual text, introductions and summaries in the form of a verbal representation or an illustration will be needed. Therefore we have *introduction* and *summary subfunctions*. The last subfunction concerns the channel through which the information is transmitted (text, illustration or both): the *channel subfunctions*. In 'formula':
Result of writer activity = f (question-description subfunction + answer-information subfunction + channel subfunction + (optional) introduction/summmary subfunction + (optional) kind of illustration).

One of the criticisms on the model of Flower & Hayes concerns the lack of clarity on the prescriptive status of the model and therefore lack of clarity in the use in instructional situations.

In our opinion a special procedure must be derived from the theoretical framework which is the starting point for constructive instructional research in the instructional practice (Terlouw, 1987). Therefore, we conceived a systematic procedure designing/writing an illustrated text based on text theory.

Question-description subfunctions
The object of description (representing information) in popular-science writing (with the emphasis on natural sciences) is mostly a:
• material *object* (apparatus or a part of it, product);
• *process* (chemical, physical, production);
• *scientific theory*.
Identifying the text theme(s) in terms of an object, process or theory offers the opportunity to develop some standard questions. These questions are often used in explanatory texts describing popular (natural-)science subjects (see Table 1).

The questions are partly interconnected; for instance, object question 4 refers to process questions. The comparison question can be connected to almost every other question and is as such very suitable for elaborating on a topic (text theme). The object and process questions 10 through 13 and the theory questions 7 through 9 can be characterised as context questions. They are not essential for the description of an object, process or theory, but they are nevertheless essential in many popular-science texts.

Defining the text themes (topic of the text and the topics of the text parts) as objects, processes or theories is one way of solving a rhetorical problem. Another way is looking at the text theme in terms of a research and/or design problem (Steehouder *et al.*, 1984). Developed from these are macrostructures such as research structure and design structure. They function as standard writing plans (see Table 2).

Table 1: Questions used in explanatory texts describing popular-science subjects

(a) Object questions	(b) Process questions	(c) Theory questions
We define an object as something with a certain physical dimension that in principle does not change in time.	In the description of a process the change in time is emphasised. There are (a) continuing and (b) discrete or batch processes charact- erised by separate stages.	1 What phenomenon does it describe?
1 What is it/What does it look like (part or whole)?		2 Which cause/effect rela- tions does it describe?
2 What attributes/charac- teristics (technical spec- ifications) does it have?	1 What is it/What does it look like (part or whole)?	3 How is the research done (methods)?
3 How is it built/What parts does it consist of?	2 What stages are there?	4 What is (are) its applica- tion(s)?
4 What is (are) its func- tion(s)?	3 What object is carrying out the (part of the) process?	5 What is (are) the pro- cess(es) the theory describes?
5 What is (are) its applica- tion(s)?	4 What is (are) its func- tions?	6 What data are known about the theory?
6 How does it work?	5 What is (are) the prod- uct(s) of (stages of) the process?	7 Who is (are) the re- searcher(s)/Where is the research done?
7 What theory is involved? 8 How is the research done (methods)?	6 What is (are) its applica- tion(s)?	8 What forms of co-oper- ation are involved?
9 How is it – industrially – produced?	7 What theory is involved? 8 What is known about (part of) the process?	9 How is it financed? 10 What can it be com- pared with?
10 By whom is it (will it be) – industrially – produced?	9 How is the research done (methods)?	
11 Who is (are) the re- searcher(s)/Where is the research done?	10 What organisation/in- dustry will use the pro- cess?	
12 What forms of co-oper- ation are involved?	11 Who is (are) the re- searcher(s)/Where is the research done?	
13 How is it financed?	12 What forms of co-oper- ation are involved?	
14 What can it be com- pared with?	13 How is it financed? 14 What can it be compar- ed with?	

In using the description questions writers should realise from their task description that the sequence presented here is not the sequence in their text scheme, c.q. in their articles. In most popular-science articles the application question, for instance, will be the first question to be answered. Writers should also realise from their task description that the extent of answering a question is not the same for all questions. For instance, the method question will hardly be answered in most cases.

A writer can, as a result of the task description, be obliged to make use of a more specific standard writing plan adapted to a specific type of periodical.

Table 2: Standard writing plans

Research structure	*Design structure*
What is the subject of the research?	What is the purpose of the design?
What is the reason for the research?	What are the requirements for the design?
What methods are used?	What means are chosen?
What are the results?	What does it look like?
What are the conclusions?	What is its value?

Answer-information subfunctions

In the section above we talked about frequently used description questions. Connected with these questions are answers in a more or less standard information structure. Especially in literature on technical writing (e.g., Lannon, 1985) and on text comprehensibility (Trimble & Trimble, 1979; Armbruster, 1984; Calfee & Curley, 1984; Horowitz, 1985a, 1985b), the (rhetorical) structures, as mentioned below, are very common.

The answer subfunctions (connected with questions: question-answer format) can be applied to information presented verbally and/or in the form of illustrations. The answer-information subfunctions are the following:

· *presentation of spatial information*. This is information about what things look like, about spatial construction, relation whole/part. Presentation of spatial information mostly occurs when the writer describes an object. Because it is very difficult to give a verbal description of a complex object in terms of its parts, one will almost certainly use an illustration here in the form of one or more photos, a cross section or an exploded view.

- *presentation of functional information.* This is information about the use of the subject. Mostly the illustration visualises an application with or without an acting person, sometimes the functions are indicated by the names of parts of an object or process.
- *presentation of attributional information.* This is information about characteristics/technical specifications of a subject. Most of the time there will be several attributes with characteristics that can be divided into two or more categories. When attributes of an object, process or scientific theory are enumerated, the most common form for an illustration is a table, but other illustrations are possible.
- *presentation of temporal information.* This is information about how something changes as time goes on. There is always a certain chronology. A temporal function is used in describing a process. The process can be the description of actions or stages but also the description of quantitative data.
- *presentation of causal information.* This concerns information about causes and effects. There can be a certain follow-up in time, but chronology is not a condition for causality. A causal information structure is used for processes or scientific theories. It can be visualised in different ways, for instance with a block diagram or with a photograph in which a cause or an effect is depicted.
- *presentation of comparative information.* This concerns information in which two (or more) things are being compared as to certain aspects (spatial construction, function, some attributes, temporal (history/old-new), causes and effects). We can distinguish between comparison in the same domain and in different domains. In the second case we speak of an analogy. In this latter form it is especially suitable for the explanation of abstract concepts (Woudstra, 1989).

Comparative information is always combined with other structures. Often these other structures are also used in combinations. For instance, most of the times the functional structure is combined with spatial structure. The attributional structure can compare spational or functional structures.

Certain answer structures are frequently coupled to certain description questions. For instance, the object questions 'What is it/What does it look like?' and 'How is it built?' always have a spatial-answer structure. The function and application questions always have a functional-answer structure in combination with a spatial or attributional structure.

Introduction and summary subfunctions

Both introduction and summary are common rhetorical structures and can be presented verbally or in the form of illustrations.

· *introduction*. An illustration acting as introduction is found at the beginning of the whole text or in a longer text at the beginning of a text part. In the illustration you often find a representation of the text theme or, if this is difficult because the topic is too abstract, an attribute or effect is visualised.

· *summary*. We talk about summarisation (by means of an illustration) when verbal information from more than one part of the text is repeated. Most of the time a visualised summarisation takes place in the form of a graphic organiser: the most important information is related graphically in a logical way.

In these two subfunctions one can, in principle, find the same answer structures as mentioned above.

Channel subfunctions

By using the channel subfunctions, the writer has the opportunity to choose between representing information in a verbal way or by means of an illustration, or by both. We distinguish:

· *duplication of information*. Verbal information is duplicated in the illustration. Of course, information in an illustration is only to a great deal identical to the information presented verbally.

· *alternation of information*. An illustration can be added to the verbal information where (part of) the information is duplicated and partly new information is presented.

· *presentation of illustrated information only*. Information is added by an illustration which has no verbal counterpart in the text. In cases where only the name of a concept is mentioned verbally and an illustration is added, we also speak of presenting illustrated information only.

· *presentation of verbal information only.* This is the kind of information we are most familiar with.

Method

Subjects

In 1989/1990 15 engineering students, all in their final year, participated in a course on popular-science writing. All students had experience with report writing, none with writing about science for a lay audience.

Materials

In the learning materials information about the text theory and the procedure was available (see Table 3).

Procedure

Data were collected during a hundred hour course on popular-science writing for engineering students in 1989/1990. We stressed the integral-function approach in the designing/writing procedure. At the end of the course students had to write an article to be published in the research periodical of the university. The articles are usually three pages including three to five illustrations. The public consists of profit and non-profit organisations, and technical and non-technical graduates from the university. The articles our students wrote were about research done by Ph.D. students. For the evaluation we gave them a questionnaire in the form of a diary in which the process of writing was followed. They had to fill in the questionnaire during the process at moments of natural rest. Other data were collected from interviews and from their articles.

Results

The results of the course were successful in terms of the written products. The editors found that 11 of the 15 articles were written well enough to be published without too much correction.

Table 3: Procedure for writing and designing illustrated texts

Task description
- Formulate the purpose of your text.
- What kind of information should the text contain?
- What are the consequences for the use of illustrations in the text hierarchy?
- What are the consequences for the type of illustration, especially realistic in proportion to more abstract diagrammatic illustrations and illustrations presenting data in graphs and tables?
- What proportion of the article space do you plan to fill with illustrations?
- Are introduction and summary subfunctions required?
- Is documentation needed?

Text scheme
- Design your text scheme in a question format.
- Formulate the answers as text subfunctions.
- Put your ideas about illustrations in the text scheme by word or sketch.
- Consult the description questions to check your text scheme plan on completeness.
- Put in new ideas for illustrations.
- If necessary, consult documentation and put new ideas for illustrations in.
- Fill in more details in your structure with more possible ideas for illustrations.

Executing phase
- Formulate your text.
- When you have difficulties in explaining a concept, use an illustration (work it out as far as you feel necessary).
- Select illustrations from your text scheme, do not work them out yet, indicate what channel subfunction you have in mind and what (combination of) other subfunctions.
- Work out your illustrations with captions (do not forget the references in your text).

'Last' evaluation (revision or edit) phase
- Check your text with illustrations with the demands of the first phase; each illustration separately and all together.
- Give a judgement in relation to the supposed effects of the purpose of the text and the conventions of the medium:
  - Is the illustration acceptable (level in text scheme, channel subfunction, other subfunctions)?
  - Are illustration and caption self-explanatory? (comprehensibility/public!, location in the text, references)?
  - Do the illustrations cause distraction because of lay-out?
  - Would other illustrations be more appropriate (level in text scheme, channel and other subfunctions, medium conventions)?
  - What would the choice for these other illustrations mean for the representation of verbal information?

But when we looked at the answers in the questionnaire the results were less positive. The planning phase caused no problems concerning the aspects related to the text conditions. In designing a text scheme ten students said they had put in ideas for illustrations, five had not. Twelve students did not consult the text theory for the use of the description questions. They did not think consultation was useful, because they already got all the information they needed from the researchers. The other three only used the questions to see whether they had forgotten anything.

In formulating the text only one student reported he had followed the instruction to formulate his actions in the form of an integrated function. The other students said the function approach was too complicated and/or not necessary because they were already satisfied with the answers (illustrations included). The (sub)functions were valued as useful only when they had their text more or less complete, ready for the last revision. Some students said the verbal presentation should be finished before thinking about channel and other subfunctions. Selection of illustrations was made at two stages: before formulating the text, as a last step in the design of a text scheme and during the 'last' revision of the text. In the revision phase the function approach and especially the use of the channel subfunction was said to be useful because it forced them to think their choices over.

Discussion

(1) We had expected the students to use the description questions to generate more information for their articles after they had interviewed the Ph.D. students about their research. However, the researchers had already provided them with enough information. In their interviews the students started with the questions from the macrostructures on research and design. The researchers also provided enough illustrations. Therefore, we should not have been surprised by the outcome.

(2) However, we were disappointed in the integrated function approach. Maybe we should pay more attention to the different writer tasks: planning, formulating and revising in relation to the utility of the different subfunctions. Because of the difficulties the students felt in using the function approach a problem solving perspective can be helpful. Until now, we only improved the linguistic aspect; the cognitive aspect will have to be improved, too.

Study two

Writing theory

Besides the linguistic criticism we also took the psychological criticism on Flower & Hayes into account. The criticism on Flower & Hayes of the psychological aspects concerns the validity of the representation of the thinking/writing processes involved. No attention has been paid to problem finding, the role of creativity, the use of schemata and the influence of different writing/designing actions (note taking, summaries, first drafts, first sketches, etc.) on the mental planning process. We will try to take this criticism into account in our procedure.

Method

Subjects
Four students from engineering departments, again in their final year before graduation, participated.

Materials
The 'Procedure for designing text schemes' was based on the adaptions we made on the Flower & Hayes model.

Procedure
Students were told to design a text scheme for an article to be published in the research periodical of the university, while thinking aloud. They had knowledge of the text theory and used the 'Procedure for designing a text schemes' (see Table 4).

Results

All four students succeeded in designing an acceptable text scheme with suggestions for illustrations. They all proceeded with step 2 after their task description. The first student of this group of four succeeded in making a text scheme of a rather global character. From the thinking-aloud protocol and from observation we noticed the following problems.

Table 4: Procedure for designing text schemes

Task description
- Formulate the purpose and subject of your text.
- What is your production scheme?
- Which is your public and what do they know about the subject?
- What are the text genre conventions concerning structure, style and lay-out?
- What are the text genre conventions concerning illustrations: kind, quantity, introductions/summaries, captions, place on the page, size, colour?
- Can you already specify your text theme and summarise what your text will look like? If not, proceed with step 1; if yes, proceed with step 2.

Text scheme
1a Brainstorm about suitable information for your text.
1b Brainstorm about suitable illustrations for your text.
1c Specify your text theme, if already possible.
1d Consult the description questions for suitable information.
1e Specify your text theme.
1f Proceed with step 2.
2 Is a macrostructure suitable, for instance the research structure or the design structure?

if yes:
- Formulate your text scheme with the help of the macrostructure questions and the function approach.
- Fill in illustrations if you think it is necessary.
- Consult the description questions if you do not know how to continue with your text scheme. Realise that you have to identify the macro-text theme (the subject of your text) and sub-text themes (the subjects of the text parts) as objects, processes or theories.
- Consult description questions to check your text scheme on completeness (do not forget the comparison question).
- Revise your text scheme considering your terms of reference.

if not:
- Consult the description questions and formulate suitable questions and answers with the help of the function approach.
- (the rest of the procedure is identical to the one formulated in '2-yes' after the first dash).

Because of the fact that the student was confronted with the topic at the beginning of the session, he had trouble in getting a global idea of the text although he could specify his text theme. It took him a lot of time to obtain such an idea. As a consequence he had trouble in taking decisions about the information in the text hierarchy and as a result the choice between object, process and theory questions sometimes confused him.

The other three students could think about their topic and the rhetorical problem during some days before the experiment. From the thinking-aloud protocols and the interviews afterward, we learned the importance of a well defined rhetorical problem for the text as a whole as formulated in the task description. They constantly reflected on their actions in a goal-oriented way. The students could work satisfactorily with the questions as part of the function approach. Their text schemes had between three and six illustrations.

However, we noticed that often they did not explicitly mention the answer-information subfunction and the channel subfunction during their text scheme task (introduction and summary subfunctions are not relevant for a writing plan). Afterwards, they reported on the answer-information subfunctions that they were sometimes so obvious that they felt no need to mention them. They were probably of more use during the actual writing of the text. They said they needed written text for an adequate use of the channel subfunction.

Discussion

Our new approach looks promising. Students need a well defined rhetorical problem; otherwise they find it difficult to make a text scheme. Students frequently consulted the description questions during their task. They could easily transform the text themes into abstractions such as object, process or theory.

Two problems require our attention. Students appreciated the functional approach because they were forced to reflect on their design actions. However, students did not use a complete function. We will probably have to simplify our function in the design procedure of the text scheme. The other problem concerns the relationship between our macrostructures and the description questions. We will have to specify this relationship in terms of hierarchy and sequence because of the fact that handling this relationship takes a lot of design time.

General discussion

(1) In our opinion a systematic procedure for text writing based on problem-solving theory can be an adequate educational tool. However, it is not quite clear how to teach it, in particular how to train and how to give feedback. Evaluation study guided by instructional design can give answers to these questions.

(2) We should be more specific about when to present information by means of an illustration. We introduced text hierarchy as a criterion but we should be more careful. It is based on the assumption that text structure is the most important aspect in the presentation of information. However, in popular-science texts attractiveness can lead to the presentation of information by means of an illustration on a very low level in the text scheme. Other criteria could be the number of words needed for the presentation of information in a verbal way, the possibility to present information in a comprehensible way and the availability of an illustration.

(3) An important point in this area concerns the extension of a caption. The extension depends on the channel subfunctions, in particular the extent in which new information has to be explained; for instance, the information added in the caption for the alternating subfunction has to be more detailed than for a duplication subfunction. The central point of interest is the *self-explanatory* character of caption and illustration together. Important aspects are: the location in the text (farther away from the place in the text where the illustration logically should be depicted means most of the time more information in the caption); the prerequisite knowledge of the audience (experts or lay readers); and the motivational clues for readers: some readers first 'read' the illustrations/captions to know whether the text is of interest to them.

(4) We have not really solved the problem of choice between text and/or illustration. This choice should be based on situational knowledge. Relevant questions to be answered are for instance:

· Which (combination of) answer subfunctions in which text situation can be represented better in text and/or illustration? In verbal representation the text is read in a chronological order. Therefore, it is not too difficult to write a well-structured text based on a combination of answer subfunctions. However, an illustration is seen as a whole. It is an interesting question how many and what kind of combinations of answer subfunctions (related to a certain public and related to the topic) still lead to a comprehensible illustration. Are some (combinations of) answer subfunctions, as for instance the causal and the comparison answer subfunction, more difficult to process than others?

· Are all aspects of the information covered by the answer subfunctions or should our text theory also contain elements as agents, actions and effects (Mosenthal & Kirsch, 1991) and information about when to add an expressive element (Peeck, 1989) or when to use a story perspective?

How to proceed with our research? For the next year we plan to continue our experiments concerning the design of a text scheme and of an actual text. We will develop a categorisation system concerning the design of the text scheme to be used in our protocol analysis. When we have tested our writing model in a satisfactory way, we will continue with the actual writing process. We will try to implement our procedure in a computer program. As a starting point we will use the SPIRIT program, developed to help high-school students in the planning phase of their writing tasks (Van der Geest, 1991).

References

Anderson, T.H., & Armbruster, B.B. (1985). Studying strategies and their implications for text book design. In T.M. Duffy & R. Waller (Eds.), *Designing Usable Texts*. New York.

Armbruster, B.B. (1984). The problem of 'inconsiderate text'. In G.G. Duffy, L.R. Roehler & J. Mason (Eds.), *Comprehension Instruction. Perspectives and Suggestions*. New York.

Calfee, R.C., & Curley, R. (1984). Structures of prose in content areas. In J. Flood (Ed.), *Understanding Reading Comprehension. Cognition, Language and the Structure of Prose*. Newark, Delaware.

Flower, L., & Hayes, J.R. (1980). The dynamics of composing: Making plans and juggling constraints. In L.W. Gregg & E. Steinberg (Eds.), *Cognitive Processes in Writing* (pp. 31-50). Hillsdale, NJ: Lawrence Erlbaum Associates.

Flower, L., & Hayes, J.R. (1981) A cognitive process theory of writing. *College Composition and Communication, 32* (4), 365-387.

Geest, Th. van der (1991). *Tools for teaching writing as a process. Design, development, implementation and evaluation of computer-assisted writing instruction.* Enschede: University of Twente (Diss).

Goel, V., & Pirolli, P. (1989). Motivating the notion of generic design within information-processing theory. The design problem space. *Artificial Intelligence Magazine, 10*(1), 19-36.

Hamel, R. (1990). *Over het denken van de architect. Een cognitief psychologische beschrijving van het ontwerpproces bij architecten.* [About the thinking process of the architect. A cognitive psychological description of the design process by architects]. Amsterdam: University of Amsterdam (Diss).

Horowitz, R. (1985a/b). Text patterns. *Journal of Reading, 28*(5), 448-454; *28*(6), 534-541.

Ikink, H. (1987). *Bekijk het eens van de andere kant.* [Have a look from another side]. Nijmegen University (Doctoral Thesis).

Lannon, J.M. (1985). *Technical writing.* Boston.

Mettes, C.T.C.W., & Gerritsma, J. (1986). *Probleemoplossen.* [Problem Solving]. Utrecht: Spectrum.

Mosenthal, P.B., & Kirsch, I.S. (1991). Mimetic documents: Process schematics. *Journal of Reading, 34*(5), 390-397.

Newell, A., & Simon, H. (1972). *Human Problem Solving.* Englewood Cliffs: Prentice.

Peeck, J. (1989). De perspectief-verlenende functies van tekstillustraties. [The perspective-supplying functions of text illustrations]. In D. Janssen & G. Verhoeven (Eds.), *Taalbeheersing in Nederland.* Groningen: Wolters Noordhoff.

Steehouder, M., Jansen, C., Maat, K., Staak, J. van der, & Woudstra, E. (1984). *Leren communiceren.* [Learning to communicate]. Groningen: Wolters Noordhoff.

Stotsky, S. (1990). On planning and writing plans – Or beware of borrowed theories! *College Composition and Communication, 41*(1), 37-58.

Terlouw, C. (1987). *De FUNDES-procedure in onderwijsontwikkeling.* [The FUNDES-procedure in instructional development]. Enschede: University of Twente (Diss).

Trimble, M.T., and Trimble, L. (1979). *The rhetoric of language for specific purposes as a model for a description of communication. Fachsprache Sonderheft 1.*

Vanmaele, L., & Lowyck, J. (1990). Het schrijfmodel van Hayes en Flower: een kritische analyse. [The writing model of Flower and Hayes: A critical analysis]. *Pedagogische Studiën, 67,* 205-221.

Woudstra, E. (1989), Analogies in non-specialist journals. *Communication & Cognition, 22*(1), 47-60.

Part II

Learner Abilities and Attitudes

7

Pragmatic knowledge in expository text comprehension: A recall study with young readers

Rachel Rimmershaw

University of Lancaster (Great Britain)

Abstract

This paper reports a study with British children between the ages of ten and twelve. The primary focus of the overall project was on learning from text (rather than recognition comprehension). In the study discussed here, young students with differing levels of background knowledge on the text topics and differing levels of pragmatic awareness in interpreting conversational interactions read two of three structurally parallel texts on different topics. It was predicted that students with good "pragmatic skills" would be more successful at making sense of difficult (i.e., topically less familiar) texts than those who had shown less pragmatic awareness in interpreting conversational contributions. The study revealed that while conversational understanding was not a significant predictor of the amount students could recall of these texts, it did play a significant role in their ability to recognise or construct the rhetorical structure of the texts, as measured by the analysis of free recalls.

Introduction

Background to the study

This research is part of a larger-scale study into learning from text focused on the processing behaviour and reading strategies of ten to twelve year olds. In Britain, children of this age are about to transfer to or have transferred to a phase of schooling in which they are expected to put their reading skills to work to learn from written materials. There is a change of emphasis from narrative to expository genres in the reading materials they are offered at school.

The understanding of narratives is underpinned by substantial assumptions of mutual knowledge between writer and reader (Gowie & Powers, 1978). This allows comprehension to proceed smoothly where these assumptions are well-founded, as it is then largely "recognition comprehension". When young students are faced with the task of learning from expository text there are also elements of recognition. These may be based on prior knowledge of the subject matter of the text (Spilich, Vesonder, Chiesi & Voss, 1979) or on recognising structural or organisational features of the text (Meyer, Brandt & Bluth, 1980; Taylor, 1980; Olson, Mack & Duffy, 1981).

In an earlier investigation of young readers' difficulties in learning from expository text (Rymaszewski, 1986) pragmatic knowledge emerged as an important resource used by these students, in addition to their knowledge of the text topic in particular and of the material and social world in general. Pragmatic knowledge is knowledge about how people get things done with language. Speakers make choices about what to say and how to say it on the assumption that their hearers will recognise the presuppositions of what is said, interpret indirect speech acts, make conversational implicatures, and so on. In interacting with a writer, readers can bring to bear the same pragmatic knowledge.

Although children's conversations seem to work on similar pragmatic bases to those analysed for adults (McTear, 1985), evidence has been reported that very young children (five to seven year olds) do not always interpret messages as attempts to represent a speaker's meaning but treat them *as* the meanings, just as they do not read intentions into actions in an adult way (Smith, 1978). In other words, they do not have an adult conception of meaning-message relationships (Olson & Hildyard, 1981; Robinson & Whittaker, 1986). This ability or preparedness to make judgements about a speaker's (or writer's) intended meaning on the basis of what they say is a measure of pragmatic knowledge. In the earlier study, readers of this age who used their pragmatic knowledge were more successful in evaluating the plausibility of probe sentences related to the text. But even the most frequent users did not make optimal use of it.

Rationale for the study

The two differences of interest in this study are differences in levels of prior knowledge, and differences in awareness of pragmatics. It was hypothesised that pupils who showed themselves adept at interpreting the participants' meanings in familiar conversational contexts would be better able to learn from a text on an unfamiliar topic by putting this knowledge to work. In other words if they treat the text as a piece of communication rather than as "objective information" they have an additional set of tools for making sense of difficult texts.

Where prior knowledge is high, new information such as unfamiliar lexical items should be readily acquired through use of specific frame properties and the semantic constraints they generate. Where prior knowledge is low, only general knowledge frames are available. Successful comprehension of new information then requires the use of a much more text-based strategy for constructing a mental representation.

This requires an assumption of textual coherence on the part of the reader, and the use of rhetorical significance cues to guide hypothesis formation. Provided they have assumed a text is coherent (the equivalent for writing of the general assumption that people's contributions to dialogues are co-operative (Grice, 1975)), readers can use the "rules" of conversational implicature, together with whatever knowledge they have of the topic and their common-sense knowledge of the world to construct the unsignalled rhetorical relations between propositions in the text.

Where pragmatic knowledge is weak, its application to the writer/reader relationship is less likely. Good "conversational competence" may or may not be spontaneously applied to text. If it is, it should be reflected in more effective processing of difficult texts.

Thus there are textual and individual variables in this study. The same text is read by students with (a) varying levels of familiarity with the topic and (b) varying levels of conversational competence. The same student reads two texts which vary in their familiarity and hence in the relative demands they place on prior domain knowledge and pragmatic knowledge.

Method

Materials

Texts
A text on the life cycle of the mosquito was selected from a children's science book (*Let's Find Out* (Vol. 1) by J. Perry Harvey) and edited to form a coherent text of a suitable length (16 sentences, 290 words – see Appendix). This text was then used as a structural template for the construction of two additional parallel texts. As far as the content allowed, the three texts were made to be syntactically and rhetorically equivalent. The rhetorical and co-reference relations in each text were determined as a basis for analysing the recall protocols. The three texts concerned:
(a)   the life cycle of a milk bottle (probably familiar to ten and twelve year olds)
(b)   the life cycle of the mosquito (possibly unfamiliar, but analogous to other known life cycles – butterflies, etc.)
(c)   the life cycle of the potato (almost certainly unfamiliar, and not easily understood by analogy).

Pre-tests and post-tests
In order to select subjects for the study for whom the texts represented appropriate contrasts of familiarity, two pre-tests were designed to establish a measure of existing knowledge in all three topic areas. A third pre-test was designed to obtain a measure of their ability to interpret conversations (by recognising presuppositions, speech acts and

indirect speech acts, judging plausibility and coherence, and drawing implicatures).

The post-tests in this study were a free recall test, a summary sentence test, and a tape-recorded interview in which nine questions were posed, and the students were required to justify their answers. This paper discusses the analysis of the free recalls.

Subjects

After piloting all the materials in another school, and modifying them where necessary, the pre-tests were administered to second and fourth year cohorts (ten and twelve year olds) in a middle school in a suburb of a northern industrial town. Thirty-three pupils in the second year and eighty-eight in the fourth year satisfactorily completed all the pre-tests. This pool of 121 pupils then had a familiarity score on each of the three topics. These were classed as "high" (60 or above), "medium" (30-59), and "low" (below 30). From this pool, the following viable contrasts emerged (see Table 1).

Table 1: Topic knowledge contrasts

| | | TOPIC | | |
	Potatoes	Mosquitoes	Milk bottles	Group
	low		med	1
	low	med		2
KNOWLEDGE	low		high	3
LEVEL		med	high	4
	low	high		5
	med	high		6
	med		high	7
	high	med		8

From the students who conformed to these profiles two pairs were selected for each group with matching scores as near as possible on topic knowledge, but divergent scores on the conversations pre-test. A third pair with an extremely low conversations score was taken from group 1, giving a total of eleven second-years and twenty-three fourth-years.

A check was done to see whether the two measures, prior knowledge and conversational understanding, were independent. This was done with all 121 subjects for whom full information was available, and these two variables were found not to be independent. Prior knowledge (scores for all three topics summed) was significantly correlated with the score on the conversations pre-test: $r = 0.59$. There was still some small degree of positive correlation between amount of topic knowledge and level of "conversational understanding" even for the 34 children in the selected sample.

Procedure

The main study was conducted two weeks after the pre-tests. The selected students were withdrawn in two groups for the first stage in which half the students read the more familiar passage and half read the less familiar passage. They were allowed to take as long as necessary until they felt they had understood all they could. When every reader was ready (after six minutes), they were asked to recall in writing as much as they could of the passage. This procedure was repeated with the second passage a few days later.

Recall protocols were analysed into propositions, which were mapped against those of the original passage to give a measure of the amount of recall, and of the proportion of relevant, text-based and veridical propositions in the protocols. "Relevance" was judged on the basis of the writer's perspective. There was between 94 and 97% agreement between codings on the number of units produced by each subject, when coding was repeated several months later. In addition to coding the units, the connections between them were coded to give measures of the extent to which the rhetorical structure had been reproduced, using the following nine categories (see Figure 1).

1	making explicit an unsignalled relation
2	reproducing a signalled relation
3	introducing a plausible and relevant new relation
4	introducing a plausible but irrelevant new relation
5	introducing an implausible new relation
<	making a new co-reference link with text-based unit
=	reproducing a co-reference link from the original surface structure
0	no explicit link made
−	no connection to a text-based unit

Figure 1: Nine categories of the rhetorical structure reproduced

Results

Although the main interest of the study is to examine whether pragmatic knowledge could play a significant role in aiding the comprehension of unfamiliar expository text, it is necessary to take account of the effects of familiarity with the text topic on the measures examined. As prior knowledge and conversational understanding were not independent, regression analyses or analyses of covariance were done in order to disentangle the effects of these two variables by taking into account the association between them.

Order and age effects

To eliminate the possibility that reading a structurally similar passage would have some kind of priming effect, raising the amount that would be recalled from the second passage, a one-tailed T-test was done on the difference between the length of the first (mean: 18.12 units) and the second recall protocol (mean: 19.62 units). The mean increase on the second passage was 1.5 units, which was not significantly different from 0 ($p = 0.35$). It was thus possible to ignore the order status of the passages in analysing the effects of familiarity and pragmatic knowledge on the post-test measures.

It was not proposed that age *per se* should be a variable of interest in this study, since any age differences could plausibly be explained in terms of the independent variables – knowledgeability about the topic and skill in understanding conversational pragmatics. So although the ten-year-old students produced fewer units in recall than the twelve year olds they also represented lower prior knowledge levels and lower levels of pragmatic knowledge, as Table 2 shows.

Table 2: Age comparisons

age	mean units in free recall	mean prior knowledge score	mean conversation score
12	20.9	48.7	15.0
10	14.7	40.4	11.1

Amount and organisation of recall by passage topics and relative familiarity

Table 3 shows the mean number of units other than irrelevant intrusions in the recall of each passage.

Table 3: Relevant units in free recall

	bottles (MB)	mosquitoes (MO)	potatoes (PO)	more familiar	less familiar
N	22	20	26	34	34
X	18.68	22.7	11.92	21.0	13.56
SD	3.71	9.12	6.24	7.44	6.63
range	13-26	8-45	3-26	12-45	3-26

These figures show that the potatoes passage was clearly least well recalled, as expected. Familiarity with the topic of the bottles passage appeared to allow a fair minimum level of recall (the highest), but the ceiling of 26 units in the recall meant that unexpectedly the mean number of units recalled was a little higher for the mosquitoes passage.

It was expected that familiarity with the topic as measured by scores on the topic pre-tests would be positively associated with free recall measures. In fact only in the case of the mosquitoes passage was prior knowledge of the topic a reasonably good predictor of recall, although on the potatoes passage the pattern was similar (but weaker). But the reverse was the case with the bottles passage, where the conversation score (cu) was a much better predictor of amount of recall than prior knowledge (pk), and significantly so. The regression equations for passage recall (number of text-based propositions) from relevant topic knowledge and conversation are shown in Table 4.

Table 4: Prediction of recall amount by passage

mosquitoes (middling)	**-4.72 +**	**0.39 mpk +**	**0.087 cu**	**r^2 = 45.7%**
t-ratios	-0.63	2.46	0.35	
significance	–	$p<.05$	–	
bottles (easy)	**8.15 +**	**0.023 bpk +**	**0.188 cu**	**r^2 = 20.3%**
t-ratios	1.52	0.26	2.02	
significance	$p<.10$	–	$p<.05$	
potatoes (difficult)	**1.8 +**	**0.187 ppk +**	**0.101 cu**	**r^2 = 22.9%**
t-ratios	0.46	1.71	0.87	
significance	–	$p<.10$	–	

The conversations pre-test scores were not significant in predicting the number of relevant general propositions produced in free recall of the less familiar passage. Prior knowledge was the significant predictor here, as the following regression equation shows.

less familiar	**7.42 –**	**0.046 cu +**	**0.300 pk (lessfam)**	**(r = 0.56)**
t-ratios	1.79	-0.4	3.42	
significance	$p<.10$	–	$p<.01$	

The regression equation for the difference scores, which takes into account the fact that each pupil read two passages, one more and one less familiar, confirms the dissociation between "conversations" score and amount recalled, and weakens the apparent effect of prior knowledge. The (non-significant) effects, however, are in the predicted directions: the greater the difference in the two prior knowledge scores the greater the decrease in amount of recall, and the greater the "conversations" score the less the decrease in amount of recall.

decrease

on less familiar 6.34 – 0.161 cu + 0.246 difference in pk

   t-ratios   1.05   -1.17   1.82

   significance –   –   –

Structure of recall

The method of judging the extent to which rhetorical structure has been perceived is by examining the rhetorical structure of the recall protocols using the number and type of intruded links (see categories in section 2.3 above). Table 5 shows how scores on the "conversations" pre-test correlated with the use of each type of link in the recall of the more familiar and less familiar texts.

These correlations show that for the less familiar passage there is a significant association between pragmatic knowledge and relevant, plausible introduced links (3), and a significant dissociation between pragmatic knowledge and using propositions which are not explicitly linked to the others even by co-reference (–).

Table 5: Correlation of "conversation" and type of link in recall

	Familiar Topic		Unfamiliar Topic	
	r	sig.	r	sig.
1	0.457	.01	0.204	–
2	0.457	.01	0.117	–
=	0.390	.05	0.303	–
3	-0.012	–	0.366	.05
4	-0.529	.01	-0.034	–
<	0.066	–	0.063	–
5	-0.017	–	-0.025	–
0	0.018	–	-0.017	–
–	-0.270	–	-0.322	.05

On the more familiar passage pragmatic knowledge is significantly associated with the reproduction of coherence ties (=) and shows highly significant associations with the reproduction of signalled links (2) and the supplying of unsignalled links (1). There is a highly significant dissociation between pragmatic knowledge and the use of plausible but irrelevant links (4).

Thus it appears that pragmatic knowledge is more closely linked with rhetorically *relevant* uses of textual links in the free recalls. This showed itself in the link-types *reproducing* the original rhetorical structure of the text quite closely (1,2,=) for the more familiar topic, and with ones *constructing* links consistent with it (3) for the less familiar topic. Such groupings of link types may also give insight into what the role of pragmatic knowledge in understanding and reproducing text structure might be, so the number of links conforming to the following groupings was calculated (see Figure 2).

		abbreviation
2+ =	reproduction of the signalled rhetorical structure	rr
1+3=	construction of a consistent rhetorical structure	cr
1+2+3+ =	maintenance of the original rhetorical structure	rm
1+2+3+4+5	signalling of connections with text-based units	sl
1+2+3+4+ < + =	use of plausible connections	pl

Figure 2: Groupings according to which the number of links was calculated

An analysis of covariance was done of how domain and pragmatic knowledge scores affected the use of the proportions of these groupings, taking also into account the relative familiarity of the texts and individual differences. The F-values are summarised in Table 6.

The analysis shows that differences between individuals and texts are not significant factors in determining these proportions, but that both prior knowledge and conversational understanding are. Prior knowledge turns out to be a significant factor on all these measures. Conversational understanding plays a smaller but still significant role in the proportion of rhetoric maintaining links produced. It is also a significant factor in the proportion of connections in the recall which are explicitly made, whereas the plausibility of the connections is most significantly affected by the level of prior knowledge.

Table 6: Signalled structure factors

	pupils df=32/32		text df=1/32		conversation df=1/32		prior knowledge df=1/32	
	F	p	F	p	F	p	F	p
rr	1.08	–	0.23	–	1.99	–	6.32	<.05
cr	0.74	–	0.035	–	1.515	–	5.30	<.05
rm	0.92	–	0.001	–	7.26	<.05	24.13	<<.01
sl	1.35	–	0.014	–	4.60	<.05	6.13	<.05
pl	1.24	–	0.53	–	3.36	–	31.58	<<.01

Discussion

Passage differences

There was a similar spread of prior knowledge scores for the milk bottles and mosquitoes topics, but high knowledge relevant to the mosquitoes passage appears to be more helpful for recall than that relevant to the bottles passage. There are grounds for believing this difference may be systematic. There is evidence that familiarity with the topic in the case of the bottles passage has lead to relative independence from the text organisation, intrusions or substitutions of the readers' own relevant knowledge, and also in some cases the adoption of a different "perspective" on the content than that signalled by the writer; all of which contribute to a lowering of the number of text-based propositions in recall. An examination of the recall protocols indicates that the "circular" macrostructure of both passages has generally been maintained, as evidenced by the fact that of the 22 students who read the bottles passage only five did not use one of the units indicating this near the end of their protocols. However, in a number of cases a more salient "cycle" schema concerning milk bottles has influenced the organisation of the recalls. As well as the writer's perspective of a cycle from the bottle's point of view, the more familiar cycle of milk delivery from the consumer's point of view has crept into some of the recalls as intrusions or elaborations. For example:

- when you have finished your milk
- the people must clean the bottles before being put out
- on a night when you put your milk bottle out
- when we have used up all the milk inside then we leave them on our doorsteps
- when the milk has been drunk

This phenomenon is most marked in recalls of the bottles passage, for which some other alternative cycles could and did interfere – the milkman's repeated journey (e.g., 118: "the milkman starts his journey again delivering milk"), the recycling of glass from damaged bottles (e.g., 23: "then they are taken to a place where they make new ones"). It is not unique to it. One or two intrusions of this type also occur in the recall of the potatoes passage, for example:

> (210) Soon the potatoes will have fully grown and soon they will be able to be took out of the ground ready for a wash.

Other types of intrusions, particularly in the potatoes passage recall, could not be accounted for in terms of an alternative perspective, but have the character of lists of any information connected with the topic object. A quotation from 22 gives the flavour.

> (22) A potato is a quite big vegetable and it is a brown or red vegetable and has been planted in a field and was thought of a long time ago.

If the particular propositions of the text to be recalled cannot easily be differentiated from other potential propositions which can be generated from activated schemata in LTM then intrusions of this type are to be expected. It has been my argument that recall for the actual propositions of a text is indexed, even when the topic is very familiar, by the rhetorical structure of the text. So those readers who draw upon the writer's explicit signals, together with their pragmatic knowledge, to recover this rhetorical structure, are less likely to produce such "irrelevant" intrusions.

Familiarity differences

It has been proposed here that adequate text comprehension requires not only the reconstruction of a semantic representation of the text content, but also of its rhetorical structure via an assumption of textual coherence. One prediction generated from this theory is that those readers who demonstrated the pragmatic skills of recognising indirect speech acts, conversational implicatures and so on in the pre-test, on the basis of a parallel assumption about felicitous contributions to conversations (Grice, 1975), should be expected to have more success in reconstructing the unsignalled rhetorical relations in a text on an unfamiliar topic than those who are less skilled. This is because they have a resource which should be more useful than topic knowledge in such circumstances, as the success rates of justifications based on conversational and lexical implicatures in an earlier study (Rymaszewski, 1986) suggest.

The advantage of using pragmatic knowledge should be reflected in performance on the free recall test, as constructing unstated rhetorical links makes more connections between propositions which should aid recall (cf. Stein & Bransford, 1979; Bradshaw & Anderson, 1982) and establishes the relevance or significance of the propositions, which should increase their chance of being incorporated into, or recoverable from, a mental macrostructure (Kintsch & Van Dijk, 1978; Van Dijk, 1980).

It is difficult to be sure how to interpret the finding that the "conversations" score was not a significant predictor of the amount recalled, but it clearly doesn't give much support to the hypothesis set out above. It is still possible that the hypothesis is not simply disproved but that perhaps the more difficult passage was in many cases so far from requiring recognition comprehension that it was beyond the possibility of learning comprehension and so led to comprehension failure. The extremely short recalls of the potatoes passage by some students suggest this may be so, although in a couple of cases their prior knowledge scores were not particularly low.

If this were so, then the hypothesis could be reformulated to predict that students reading passages which were judged to be moderately difficult for them on the basis of the pre-tests would be most helped by pragmatic knowledge. But the results do not support this version of the hypothesis either. When the recall scores for the moderately difficult passage (i.e., pre-test scores between 30 and 59, n = 28) were predicted from prior knowledge and the conversations pre-test score, prior knowledge was still the only significant predictor ($p < .05$ on a one-tail test).

The regression equation is

recall =	**0.66 +**	**0.092 cu +**	**0.288 pk**
t-ratios	0.09	0.81	1.87
significance	–	–	$p < .10$

These results appear to indicate that in this sample of ten to twelve year olds, pragmatic awareness in conversational contexts does not transfer to textual contexts in a way which affects the amount of the text that is recalled.

Structure of recall

Perception of the writer's rhetorical structure is likely to be reflected in the presence of intruded links which make explicit unstated rhetorical relations of the original text, and recall of high-level propositions. The smaller the reader's existing domain knowledge base, the more tentative such constructions will be, but a reader who does not apply the rules of conversational implicature (especially the maxims of quality and relation) will often fail to perceive unsignalled rhetorical links at all. One would expect such failures to be reflected in how the recall is organised. The protocols of those who do pick up the writer's intentions are more likely to include intruded links (making explicit unsignalled rhetorical relations) and inferences (supporting hypothesised significances). If the reader does not pick up the significances the writer intended they may:

(a)   omit apparently irrelevant propositions or their components from free recall (hence omissions and overgeneralisations)

(b)   reorganise the rhetorical structure of the passage to reflect a script or frame which the topic or some aspect of it has activated in their own knowledge base (hence intrusions, intruded links and signals, reordering of propositions, omissions, overelaborations)

(c)   treat the passage propositions as unrelated chunks of information (hence list-learning phenomena, such as primacy and recency effects, absence of connections).

The results presented in this paper are consistent with this interpretation of readers' behaviour. So although conversational understanding does not appear to affect the *amount* of recall, it does seem to play a part in affecting qualitative aspects of the structure of the recalls, being positively associated with explicit links which maintain the writer's original rhetoric. This gives some support to the arguments for a role for pragmatic knowledge in constructing a mental representation of the rhetorical structure when comprehending a text. However, the argument that this phenomenon should be sensitive to the relative unfamiliarity of the text is not straightforwardly supported.

Summary

As expected, the familiarity of topic as measured by the prior knowledge pre-test scores had an effect on all the measures examined here. The question of interest concerned the extent to which pragmatic skills, as measured in a more conversational context, might play a part in the successful comprehension of expository text by readers of this age.

Such skill was found not to be related to measures of the amount recalled, though some caution is needed in interpreting this finding. An examination of the recall protocols revealed that prior knowledge of the two passages (mosquitoes and bottles) with a similar spread of scores was used in differing ways. Knowledge of life cycles analogous to the mosquito's was used consistently with the passage rhetoric, whereas knowledge of other perspectives on the bottles cycle was sometimes used to switch to other, perhaps more familiar perspectives, to render irrelevant some of the original passage content, and thus to reduce the number of text-based propositions in the recall.

Conversational understanding scores were found to be related to measures of the maintenance of the rhetorical structure in the free recalls. So it is suggested that the ability to interpret communicative intentions, while not significantly influencing the number of propositions a reader learns, may influence how those ideas are organised. These general findings, however, mask considerable individual differences in the use of pragmatic knowledge, which will be the subject of future publications from this research.

References

Bradshaw, G.H., & Anderson, J.R. (1982). Elaborative encoding as an explanation of levels of processing. *Journal of Verbal Learning and Verbal Behavior, 21*, 165-174.

Gowie, C.J., & Powers, J.E. (1978). Children's use of expectations as a source of information in language comprehension. *Journal of Experimental Child Psychology, 26*, 472-488.

Grice, H. (1975). Logic and conversation. In P. Cole & J. Morgan (Eds.), *Syntax and Semantics, Vol. 3: Speech Acts*. New York: Academic Press.

Harvey, J.P. (1959). *Let's Find Out, Vol. 1*. London: Evans Bros.

Kintsch, W. & Van Dijk, T.A. (1978). Towards a model of text comprehension and production. *Psychological Review, 85*(5), 363-393.

McTear, M.F. (1985). *Children's Conversations*. Oxford: Basil Blackwell.

Meyer, B.J.F., Brandt, D.M., & Bluth, G.J. (1980). Use of top-level structure in text: Key for reading comprehension of ninth-grade students. *Reading Research Quarterly, 16*, 72-103.

Olson, D.R., & Hildyard, A. (1981). Assent and compliance in children's language. In W.P. Dickson (Ed.), *Children's oral communication skills*. New York: Academic Press.

Olson, G.M., Mack, R.L., & Duffy, S.A. (1981). *Cognitive aspects of genre*. Cognitive Science Technical Report No. 11, University of Michigan.

Robinson, E.J., & Whittaker, S.J. (1986). Children's conceptions of meaning-message relationships. *Cognition, 22*, 41-60.

Rymaszewski, R.H. (1986). *Coherence and rhetoric in children's text comprehension*. CeRCLe Technical Report No. 18, Lancaster University.

Smith, M.C. (1978). Cognizing the behaviour stream: The recognition of intentional action. *Child Development, 49*, 736-743.

Spilich, G.J., Vesonder, G.T., Chiesi, H.L., & Voss, J.F. (1979). Text processing of domain-related information for individuals with high and low domain knowledge. *Journal of Verbal Learning and Verbal Behaviour, 18*, 275-290.

Stein, G., & Bransford, J.D. (1979). Constraints on elaboration: Effects of precision and subject generation. *Journal of Verbal Learning and Verbal Behaviour, 18*, 767-777.

Taylor, B.M. (1980). Children's memory for expository text after reading. *Reading Research Quarterly*, *3*, 399-411.

Van Dijk, T.A. (1980). *Macrostructures*. Hillsdale, NJ: Lawrence Erlbaum Associates.

Appendix

Base passage (medium familiarity): MO

The mosquito is one of those insects which lead a double life. We know them as flying insects which may bite us on a summer evening, but the first stages of the mosquito's life take place out of sight in quite another world.

Mosquitoes lay their eggs in the still water of swamps and ponds. When the eggs hatch, a tiny larva or "wroggler" comes out. The larva lives in the water for a few days until it has grown up into an adult mosquito. It has tiny tubes at the end of its tail for breathing, but it cannot breathe in the water. The larva has to come up to the surface and poke its tail into the air. Then it can swim down to the bottom of the pond again to feed. It stirs up the mud with its mouthparts to find edible matter. Later on the larva turns into a comma-shaped pupa, which no longer feeds but floats at the surface of the water until the time comes for the adult to emerge. At this moment the pupa case splits open. After resting for a while on the empty pupa case the adult flies off to begin the last stage of its life.

During the flying period of its life the male mosquito does not feed, but the female, who will produce the eggs, needs extra food. When a female mosquito "bites" an animal it actually pierces the skin with a fine tube called the proboscis. Through this tube it sucks up a supply of blood before flying off to find another meal. Once the mosquitoes have mated, and the female has laid her eggs in a suitable place, the life cycle begins again.

8

Comprehension of story content in adolescents: Influences of age and cognitive competence[1]

Christiane Spiel

University of Vienna (Austria)

Abstract

The study is part of a research programme focusing on children's and adolescents' competence in text comprehension. The current experiment was designed to investigate whether the performance in story comprehension – coherent understanding of an interpersonal conflict – depends on cognitive development, i.e., on the adolescents' ability to discover formal operations. Several studies have demonstrated that performance in comprehending stories improves with age. But there is still a great variation in the age at which subjects achieve the level of formal operations. Seventy fifth to eighth-grade students were asked to summarise the main ideas of a story. Additionally, a syllogism test was employed to test their cognitive level. Results indicate that both conditional reasoning ability and understanding of interpersonal conflicts are indicators of the same latent developmental process.

Introduction

Text comprehension is assumed to depend on two general aspects: structure and content. The structural aspect of text comprehension is

---

[1] I would like to thank S. Diwald and V. Kranewitter for the data collection.

assumed to be controlled by specific schemata that help us to identify a particular discourse as an example of a general one. The term 'story schema' refers to an idealised internal representation of the components of a typical story and the relationships among those parts. Various grammatical analyses of story texts have been suggested to describe the components of a story (e.g., Rummelhart, 1985; Schank & Abelson, 1977; Stein & Glenn, 1979; Thorndyke, 1977). The content aspect concerns the process of text comprehension. Different theories were formulated describing the process of text comprehension while reading a text (e.g., Collins, Brown & Larkin, 1980; Van Dijk & Kintsch, 1983; Johnson-Laird, 1980; Kintsch & Van Dijk, 1978; Mandl & Schnotz, 1985).

Studies exploring the development of children's and adolescents' knowledge about stories show that performance in comprehending story structure and content improves with age. Results demonstrate that certain components of stories are well-recalled by first-graders; for example, setting, initiating events, and consequences (Mandler & Johnson, 1977; Nezworski, Stein & Trabasso, 1982). Fitzgerald, Spiegel & Webb (1985) found a developmental increase from the fourth grade to the sixth grade in the occurrence of internal conflicts in a story production task. Ackerman's (1988) results concerning story content show developmental differences in the ability to make reasoning inferences, which in his opinion have to do with conceptual knowledge. In addition, younger children showed substantial deficiencies in inference modification. However, the age-related improvement on text comprehension in children and adolescents is caused by developmental changes in cognitive competence. But there is still a great variation in the cognitive competence of subjects of the same age (Piaget, 1968). Nevertheless, most studies analysing text comprehension in children or adolescents focus on age differences and not on differences in cognitive competence.

Previous experiments

In the present research programme, a story with two levels, "Hannes fehlt" ("Hannes is missing") by Ursula Wölfel (1982), is used. In the story the student Hannes is lost during a school trip. After a while he is found by the teacher (Level 1). In the meantime Hannes' classmates realised that they did not know anything about Hannes and had never really noticed Hannes before (Level 2). Level 1 is actional and fully fits a story schema. Thus, it is possible to recall only Level 1 of the story without losing any coherence. In Level 2, there is a conflict in interpersonal relations described which must be inferred from the dialogues of the students. It is not possible to recall only Level 2 without losing coherence. Fifteen psychology students rated the main theme of the story. There was total agreement that the main theme is contained in Level 2.

A first experiment (Spiel, 1987) was done to investigate whether adolescents, in summarising the story, focused on the interpersonal conflict (Level 2) in the same way the psychology students did. Seventh-grade students ($n = 108$) were instructed to summarise the story "Hannes is missing". Level 1 was recalled by 92% of the subjects in a coherent way, which means that the story structure was known to these subjects. Only 19.4% of the subjects recalled Level 2 in a coherent way.

Several studies have demonstrated that a thematic title which grasps the general message of a story facilitates comprehension because it permits the reader to concentrate on the encoding of the main ideas (Anderson & Pirchert, 1978; Anderson, Pirchert & Shirey, 1983; Kozminsky, 1977; Pirchert & Anderson, 1977; Schwarz & Flammer, 1981). Thus, a second experiment was done (Spiel, 1990) with 78 sixth-graders to investigate whether a title that focused on the interpersonal conflict would raise coherent recall of Level 2 of the story. Results confirmed that assumption. The title focusing on the interpersonal conflict significantly raised the coherence in the recall of Level 2 of the story.

Aims of the current study

The goal of a third study was to investigate the influence of age and cognitive competence on comprehension of story content. Based on results of previous studies a linear age-related increase was expected. In addition, the experiment investigated whether performance in story comprehension depends on cognitive development, that is, on the child's ability to discover formal operations (Piaget, 1969). According to Piaget, the stage of formal operations implies the ability to formulate and test hypotheses, and to detach the concrete logic from the objects themselves. In his opinion, the intellectual changes in adolescent cognition occur not only in the ability to build and understand abstract theories: the adolescent is also capable of taking a multiple perspective. Piaget maintained that during adolescence, parallel to cognitive development, egocentrism and narcissism decrease (Piaget, 1969). This process seems to be a basic premise for understanding interpersonal conflicts; therefore, it was expected that subjects who were better able to discover and solve abstract operations would also be better able to discover and understand an interpersonal conflict. This cognitive developmental process starts between the ages of 11 and 12 (Piaget, 1968) and does not finish until 14 or 15, although there is still great variation in the age at which adolescents discover formal operations. According to Piaget, some individuals remain at the level of concrete operations throughout their entire life. Therefore, not only the age of the subjects, but also their cognitive level was expected to have an effect on the comprehension of the story content. There is a lack of studies which compare story comprehension and cognitive competence.

In the present experiment fifth to eighth-grade students were asked to summarise the story "Hannes is missing". In addition, a conditional reasoning test was used to assess their cognitive development. It was assumed that coherent understanding of the interpersonal conflict would depend on both grade and cognitive level. Sex differences were not investigated because no sex differences were observed in a previous experiment (Spiel, 1990).

Method

Subjects

21 fifth-graders, 23 sixth-graders, 13 seventh-graders and 13 eighth-graders from two schools in Vienna participated in the study.

Material

To measure story comprehension, the two-level story "Hannes is missing" was used. To develop a scoring system, the story was first summarised independently by three psychologists and then combined into a final version containing ten criterion ideas, five for Level 1 and the other five for Level 2. Similar to Schwarz & Sinninger (1984), a criterion idea is defined as a scene which comprises a central action. In addition, fifteen psychology students rated each criterion idea as to whether it belonged to Level 1 or to Level 2. There was full agreement between the ratings. To measure the cognitive level, a conditional reasoning test was used. This test is an extended version from the conditional reasoning test developed by Kodroff & Roberge (1975) and revised by Schröder (1989). The test contains two syllogistic tasks, for each of which four basic forms of syllogistic arguments were presented: (1) affirmation of antecedent, (2) denial of antecedent, (3) affirmation of consequent, and (4) denial of consequent. The first task refers to a real life situation; in the second task an abstract content is presented. The number of correct answers was used as a measure of conditional reasoning ability. The eight items were shown to be homogeneous as defined by the dichotomous logistic model (Rasch, 1960). Therefore, the summary scores are sufficient estimators for the subject's competence.

Procedure

The experiment was conducted during regular class instruction, and the subjects did not know that they were participating in a study. The duration was approximately 45 minutes. The subjects were required to summarise the story in about ten sentences and to do the conditional reasoning test. For each level of the story, the number of criterion ideas of the experts' summary contained in the subjects summaries were counted by two in-dependent judges. The agreement between the two judges was obtained by computing the value of kappa (Cohen, 1960). The resulting value was .96. In addition, both levels were scored for coherence, that is, whether the levels were not recalled, incoherently recalled, or coherently recalled. The inter-rater agreement was 100%. For the conditional reasoning test, the number of correct answers was counted.

Design

A single-factor design was used, with grades as the independent variable. The dependent variables were the number of criterion ideas contained in the subjects' summaries for both story levels, the coherence ratings for the story levels, and the conditional reasoning test scores.

Results

The results showed significant grade effects for the number of Level 2 criterion ideas contained in the subjects' summaries ($F(3,66)=5.98$, $p<.05$). Duncan analyses demonstrated that the summaries of eighth-grade students contained significantly more Level 2 criterion ideas than the summaries of all lower-grade students. There were no differences between fifth-, sixth-, and seventh-grade students. The means and the standard deviations are shown in Table 1. There were no significant grade differences in the number of Level 1 criterion ideas contained in the students' summaries.

Significant grade differences were observed in the conditional reasoning test scores ($F(3,66)=7.18$, $p<.05$). Duncan analyses showed that eighth-grade students gave significantly more correct answers than the lower-grade students. There were no differences between the fifth-, sixth- and seventh-grade students. The means and the standard deviations are shown in Table 1.

Table 1: Means and standard deviations of the conditional reasoning test scores and of Level 1 and Level 2 criterion ideas contained in the subjects' summaries for each grade

Grade	Test Scores		Level 1 Criterion Ideas		Level 2 Criterion Ideas	
	mean	SD	mean	SD	mean	SD
Fifth	5.48	1.33	4.33	0.66	2.71	0.72
Sixth	4.83	0.83	4.61	0.58	2.78	0.95
Seventh	5.38	0.87	4.77	0.44	2.76	1.30
Eighth	6.54	1.13	4.38	0.65	3.92	0.49

Table 2: Distribution (%) of Level 2 recall coherency by grade in school

Grade	No Recall	Incoherent Recall	Coherent Recall
Fifth	0.0	90.5	9.5
Sixth	4.3	82.6	13.1
Seventh	15.4	46.1	38.5
Eighth	0.0	15.4	84.6

To compare the grades on recall coherence, Chi-square analyses were computed. There were significant differences in the coherence ratings (coherent versus incoherent recall) of Level 2 (Chi-square $= 26.01$, df $= 3$, p $< .05$). Students from the higher grades coherently recalled Level 2 of the story more often than those in the lower grades (see Table 2). There were no grade differences in the coherence ratings for Level 1. 91.4% of the subjects recalled Level 1 coherently.

To analyse the relations between grade, conditional reasoning ability, and comprehension of story content, correlation coefficients were computed (see Table 3). While the correlations of Level 1 coherence ratings and Level 1 criterion ideas with test scores and grade level were not significant, the correlations of Level 2 coherence ratings and Level 2 criterion ideas with test scores and grade level did reach significance. Furthermore, negative correlations were found between Level 1 and

Level 2 both in coherence ratings and the number of criterion ideas. The correlation between coherence ratings and number of criterion ideas was higher for Level 2 than for Level 1.

Table 3: *Intercorrelations between conditional reasoning test scores, number of criterion ideas for each level contained in the subjects' summaries, coherence ratings for each level, and school grade level (* p<.05)*

	Test Scores	Level 1 Criterion Ideas	Level 1 Coherence Ratings	Level 2 Criterion Ideas	Level 2 Coherence Ratings
Level 1 Criterion Ideas	.00				
Level 1 Coherence Ratings	.11	.43*			
Level 2 Criterion Ideas	.21*	-.29*	-.01		
Level 2 Coherence Ratings	.27*	-.15	-.14	.76*	
Grade	.30*	.08	-.02	.37*	.46*

To investigate the effect of conditional reasoning ability on comprehension of story content, adjusted for grade effect, two partial correlation coefficients were computed. The partial correlation between the number of Level 2 criterion ideas contained in subject's summaries and the conditional reasoning test score was .11 (p=.19). The partial correlation coefficient between the test score and Level 2 coherence rating was .16 (p=.10).

Discussion

The results show that story structure (Level 1) was well known in all four grades investigated. The developmental process underlying a child's ability to identify a story's structural components seems to be completed before the age of eleven. The hypothesis that the coherent understanding of the interpersonal conflict (Level 2) is age-related was supported by the results. The eighth-grade students recalled Level 2 quantitatively (number of criterion ideas) and qualitatively (coherence of recall) better than the lower-grade students. A somewhat surprising finding was that there were no differences between fifth, sixth and seventh-graders. Thus the assumption concerning a linear age-related increase in comprehension of story content could not be confirmed. The findings indicate that the ability to discover an interpersonal conflict in a story increases abruptly. The results concerning conditional reasoning ability show a similar picture. These observations indicate that the developmental process underlying conditional reasoning ability is also transformed abruptly. This parallel between conditional reasoning ability and coherent understanding of the interpersonal conflict supports Piaget's (1969) assumption that the decrease in egocentrism during adolescence, parallel to cognitive development, is a basic premise for understanding interpersonal conflicts. The modest intercorrelations between test scores and Level 2 summary scores, syllogism scores, and Level 2 coherence scores may be due to the lack of differences between fifth, sixth and seventh-graders. According to the results, conditional reasoning ability had no effect on understanding the interpersonal conflict when adjustments were made for grade effect. These findings support the conclusion that both conditional reasoning ability and understanding of interpersonal conflicts are indicators of the same latent developmental process. Another interesting finding was the tendency towards negative correlations between the recall of Level 1 and Level 2, both in number of criterion ideas and in coherence of recall. One explanation could be that when the subjects focused on Level 2 of the story they tended to neglect Level 1 and vice versa.

134

References

Ackerman, B.P. (1988). Reason inferences in the story comprehension of children and adults. *Child Development, 59*, 1426-1442.

Anderson, R.C., & Pirchert, J.W. (1978). Recall of previously unrecallable information following a shift in perspective. *Journal of Verbal Learning and Verbal Behavior, 17*, 1-12.

Anderson, R.C., Pirchert, J.W., & Shirey, L.L. (1983). Effect of the reader's schema at different points in time. *Journal of Educational Psychology, 75*, 271-279.

Cohen, J.A. (1960). A coefficient of agreement for nominal scales. *Educational and Psychological Measurement, 20*, 37-46.

Collins, A., Brown, J.S., & Larkin, K.M. (1980). Inference in text understanding. In R.J. Spiro, B.C. Bruce & W.F. Brewer (Eds.), *Theoretical issues in reading comprehension*. Hillsdale, NJ: Lawrence Erlbaum Associates.

Van Dijk, T.A., & Kintsch, W. (1983). *Strategies of discourse comprehension*. New York: Academic Press.

Fitzgerald, J., Spiegel, D.L., & Webb, T.B. (1985). Development of children's knowledge of story structure and content. *Journal of Educational Research, 79*, 101-108.

Johnson-Laird, P.N. (1980). Mental models in cognitive science. *Cognitive Science, 4*, 71-115.

Kintsch, W., & Van Dijk, T.A. (1978). Towards a model of text comprehension and production. *Psychological Review, 85*, 363-394.

Kodroff, J.K., & Roberge, J.J. (1975). Developmental analysis of the conditional reasoning abilities of primary-grade children. *Developmental Psychology, 11*, 21-28.

Kozminsky, E. (1977). Altering comprehension: The effect of biasing titles on text comprehension. *Memory and Cognition, 5*, 482-490.

Mandl, H., & Schnotz, W. (1985). *New directions in text processing*. Deutsches Institut für Fernstudien an der Universität Tübingen, Forschungsbericht 36.

Mandler, J.M., & Johnson, N.S. (1977). Remembrance of things parsed: Story structure and recall. *Cognitive Psychology, 9*, 111-151.

Nezworski, T., Stein, N.L., & Trabasso, T. (1982). Story structure versus content in children's recall. *Journal of Verbal Learning and Verbal Behavior, 21*, 196-206.

Piaget, J. (1968). *Judgment and reasoning in the child*. Towota, NJ: Littlefield, Adams.

Piaget, J. (1969). The intellectual development of the adolescent. In G. Caplan & S. Lebovici (Eds.), *Adolescence: Psychological perspectives* (pp. 22-26). New York: Basic Books, Inc.

Pirchert, J.W., & Anderson, R.C. (1977). Taking different perspectives on a story. *Journal of Educational Psychology, 69*, 309-315.

Rasch, G. (1960). *Probabilistic models for some intelligence and attainment test*. Kopenhagen: The Danish Institute for Educational Research.

Rummelhart, D.E. (1975). Notes on a schema for stories. In D. G. Bobrow & A. M. Collins (Eds.), *Representation and understanding: Studies in cognitive science*. New York: Academic Press.

Schank, R., & Abelson, R. (1977). *Scripts, plans, and understanding*. Hillsdale, NJ: Lawrence Erlbaum Associates.

Schröder, E. (1989). *Vom konkreten zum formalen Denken. Individuelle Entwicklungsverläufe von der Kindheit zum Jugendalter*. Bern: Huber.

Schwarz, M.N.K., & Flammer, A. (1981). Text structure and title effects on comprehension and recall. *Journal of Verbal Learning and Verbal Behavior, 20,* 61-66.

Spiel, C. (1987). *Influence of learning on comprehension and summarization of stories – a field study.* Paper presented at the Second European Conference for Research on Learning and Instruction, Tübingen.

Spiel, C. (1990). Zum Einfluß von Textinhalt, Titel und Vorwissen auf das Textverstehen. *Zeitschrift für Experimentelle und Angewandte Psychologie, 37,* 505-518.

Stein, N.L., & Glenn, C.G. (1979). An analysis of story comprehension in elementaristic school children. In R. Freedle (Ed.), *New directions in discourse processing* (Vol. 2). Hillsdale, NJ: Ablex.

Thorndyke, P.W. (1977). Cognitive structures in comprehension and memory of narrative discourse. *Cognitive Psychology, 9,* 77-110.

Wölfel, U. (1982). *Die grauen und die grünen Felder.* Ravensburg: Ravenburger.

9

Affect, reading behaviour and reading achievement

Huub van den Bergh
University of Utrecht (The Netherlands)
Gert Rijlaarsdam
University of Amsterdam (The Netherlands)

Abstract

In this study we reanalysed the data of a national assessment on language skills in the Netherlands with respect to the relations between attitudes and reading achievement. This gave rise to two specific questions. First, is the relation between attitudes and reading achievement invariant over schools? And secondly, is the relation between the behavioural component and reading achievement invariant over different subpopulations (achievement levels)? It was concluded that the relations between attitude, behaviour and achievement seem to depend on the specific operationalisation of the instruments for measuring reading achievements, but these relations do not seem to vary across schools. The relationship between behaviour and achievement proved to depend on the level of proficiency.

Introduction

To influence and enhance the affect related to and to stimulate language activities are educational goals. In the mother tongue curriculum these goals are considered important. They are advocated in handbooks for teachers (Kahn & Weiss, 1973; Leidse werkgroep moedertaaldidactiek, 1986), but also practised (see for instance Hoogeveen & Verkampen (1985) and Bos & Oostdam (1985) in the Netherlands).

Attitudes can be conceived of as consisting of, among other things, an affective and a behaviourial component (Rosenberg & Hovland, 1960). Teachers try to influence both components. In the case of reading instruction the stimulation of the affective component can be thought of as encouragement to read, to enhance the joy of reading, but also as the change of negative feelings towards reading and books. The behaviourial component might be indicated with the encouragement of the use of the library, to enhance free-time reading and to get the students acquainted with 'good' books.

Sometimes, the legitimation of the attitude goals is stated in non-causal relationship to language skills. Teachers who stress attitudinal goals for their own sake argue that the willingness to read is an important goal next to the development of reading ability. In their opinion there is no use in being a skilled reader if one does not like reading and therefore will not read (Van den Bergh & Kuhlemeier, 1988). Thus, this insight would result in a reading programme that stresses book promotion, talking about books and writers in the class etcetera. On the other hand, there are teachers who reason the other way around. In their opinion reading skill programmes must be stressed and the activities to enhance the willingness to read are merely a derivative. But mostly, some causal relationship between attitudes and achievement, in whatever direction, is presupposed, either explicitly or implicitly. The direction mostly endorsed is a direction from attitudes to achievement: the stimulation of positive and the alteration of negative attitudes to and fears of language activities and language behaviour have been expected to have a positive effect on language skills (Mitzel, 1982). Students not willing to read will have lower scores on reading tests than their peers who like and enjoy reading (Fishbein, 1967; Wicker, 1969; Brett, 1978; Condon, 1978; Cialdini, Petty & Cacioppo, 1981).

In studies concerning the relation between attitudes and achievement this relation is (implicitly) formulated at the student level: *students* who have a high ... will have higher reading scores than *students* ... But in all these studies the respondents, viz. students, are nested within schools. Hence, two variance components have to be distinguished: a between-school component and a between-student component. If one does not take this hierarchical structure into account, one is bound to make mistakes (Cronbach, 1979, p. 1), even in carefully planned experimental conditions (Bryk & Raudenbush, 1981, p. 66).

There are also theoretical reasons for distinguishing variance components at different levels. One might expect that teachers differ in the amount and type of activities meant to enhance the attitudes towards reading and writing. Therefore it is interesting to study the covariances between these activities and achievements at the school level. Besides, one might also expect that the same activity does not have the same effect on each school (Prosser, Rasbash & Goldstein, 1990). This can only be investigated if the variances of different levels are separated (Aitken & Longford, 1986).

A second problem concerns the linearity of the relationship between attitudes and achievements. It is mostly assumed that the relationship is linear. But, for instance, the relation between television watching and academic achievement has been shown to be non-linear (Ward, Mead & Searls, 1983). For low-ability students there was a negative correlation, whereas for the others a positive correlation was revealed. This effect could also show up for time engaged in reading. In that case the explanation might go like this: high-ability readers read books of a relative high level of difficulty (language use, theme, structure), and low-ability readers read, with the same amount of pleasure, pulps and comics. Both subpopulations spend the same amount of time reading. The low-ability readers, however, are missing the opportunity to learn from reading: more reading of low-level books does not influence their reading ability. Median and high-ability readers might give themselves the opportunity to learn. If this is true, one could only expect a linear relation within subpopulations of mediocre and high achievers.

In this article we will pose two questions. The first question has to do with the distinction between student and school level: Is the relation between the affective and behaviourial component on the one hand and achievement on the other hand invariant across schools?

The second questions concerns the differences between subpopulations: Is the relation between the behaviourial component and reading achievement invariant across different achievement levels?

Method

Subjects

Subjects were 2147 sixth-grade students (aged about 11), from 154 schools who took part in the national assessment on language skills (Wesdorp, Van den Bergh, Bos, Hoeksma, Oostdam, Scheerens & Triesscheijn, 1986).

Instruments

Sixteen tasks were used to assess reading comprehension. As usual in assessments some type of matrix sampling was used: not all students took all tasks. The tasks were organised in thematically related booklets consisting of both reading and writing tasks. Every student took only one of the booklets.

For instance in a booklet 'Journey to Waddenoog', a non-existent island, the student helps organise a school trip to an island. He writes a letter to the tourist office in which information is requested about the youth hostels on the island. The tourist office returns a brochure about the island. In this brochure the student can find information about the youth hostels. This information is given in the form of several tables. The student has to choose a youth hostel which meets several restrictions, like being open out of the tourist season, costs, etcetera (reading 1). On the basis of a letter from the owner of a youth hostel and the reaction from the school the student has to fill in a form (reading 2). Of course, the student has to take note of the possibilities and the wishes as expressed in both letters. Next, the journey has to be planned. In order to do that, timetables of the departures and arrivals of buses and boats have to be studied (reading 3). One of the texts in the brochure is about the dunes of the island. The student has to read the text and answer some questions (reading 4) in order to prepare himself to tell something about the dunes to his classmates. Once on the island, the student has to describe the youth hostel for the school paper. In order to do that, he has to use the illustrations in the brochure. He also has to write a letter with a route description from the bus stop to the hostel for one of the teachers who will arrive later. For this, the student has to use the map of the island, which is given in the brochure. Next, he has to compare two texts about two musea on the island in order to decide which one will be visited (reading 5). On his last day on the island, the student has to write an essay about his mood on a rainy day when he is alone.

The other two booklets were similar in the way that different reading and writing tasks were combined in a thematic way, following the scenario principle. In Table 1 some information is given about the reading scores on the three booklets.

Table 1: Mean (X), standard deviations (SD), number of reading tasks (N_{tasks}), number of students ($N_{students}$), number of schools ($N_{schools}$) and a reliability estimate (α) per booklet

Booklet	X	SD	N_{tasks}	$N_{students}$	$N_{schools}$	α
A	40.16	8.31	7	566	153	.88
B	41.21	5.12	5	594	154	.84
C	27.63	6.23	5	577	153	.87

Attitude was measured with a questionnaire, containing different subscales. The subscales can be grouped into behaviourial and affective ones. In Table 2 we give a summary of the aspects related to reading which were covered with the questionnaire, and an example of the questions for each aspect. Next to the attitudinal aspects, information was gathered about several background features: gender, language spoken at home, and type of school as a proxy-variable of social economic status. Information on television watching is taken into account, too, because this is competitive spare time behaviour: if one watches television, one can not read.

As can be seen in Table 2 six aspects related to reading affect or reading behaviour are measured. The students who took the reading tasks also responded to the questionnaire. Therefore, we are able to crossvalidate the findings for each booklet.

Table 2: A summary of the aspects of reading attitude and reading behaviour

Variable	Description
Affective component	
Comics	Favouring books over comics
Reading preference	Types of books a student likes to read
Reading process	Response to items like: 'I have read a book for the second time, because I liked the book so much', 'Sometimes reading is so fascinating that I forget everything'
Behaviourial component	
Reading time	Time spent on reading after school time (expressed in quarters of an hour)
Reading behaviour	Types of books a student reports to have read during the last month
TV-time	Averaged number of minutes spent watching television
Background variables	
Stimulation school	A school variable indicating whether there are many children with deprived backgrounds visiting the school
Gender	Boys versus girls
Language at home	Language spoken at home: Dutch or other

Results

In the analyses we use all nine independent variables as measured – eight at the individual level and one at the school level.

In the first fitted model all nine individual variables were fixed at the student level. Only the intercept was allowed to vary across schools; for each school a separate intercept was estimated. We can write the model for student i at school j for subscale k ($k = 1, 2, ..., 9$) as:

$$Y_{ij} = \beta_{0j} + \Sigma_k \beta_k X_k + e_{ij}$$

$$\beta_{0j} = \gamma_{00} + \mu_{0j}, \tag{1}$$

Or in one equation as:

$$Y_{ij} = \gamma_{00} + \Sigma_k \beta_k X_k + (e_{ij} + \mu_{0j}) \tag{2}$$

which looks just like a traditional regression equation, with only a more complex residual term. That is, there are residuals at student (e_{ij}) as well at school level (μ_{0j}). We therefore have specified a genuinely multi-level model.

Next, all independent variables which did not contribute to the explanation of the differences in reading achievement were removed from the model. The resulting model was refitted. In the following models one by one all slopes were either random or fixed. If a slope contributed significantly to the variation between schools, it remained random in the following model. Next, several interactions were tried; interactions between gender and reading time, gender and reading behaviour, etcetera. Note that all models were fitted thrice: once for each booklet. It appeared that none of the parameters at the fixed level contributed significantly to the differences between schools. That is, on the basis of these independent variables we were not able to explain differences in reading achievement between schools, other than the differences between students; the effect of student activities related to reading is invariant across schools.

For the second research question we have to define interactions between achievement level and behaviourial scores. We limited ourselves to the low achievers (i.e., students with a score of at least one standard deviation below the mean score). For these students an extra parameter for all three behaviourial components (in all three booklets) was estimated. Only the interactions which reached significance are presented in Table 3: the effect of reading time and TV-time for low achievers only.

None of the interaction effects tested – gender x reading time, language at home x t-time, etcetera – did contribute to the explanation of reading achievement. We are therefore left with our nine initial and two constructed variables. The results are summarised in Table 3.

As can be seen in Table 3 the regression weights for all eight main effects are relatively small, or completely absent. For instance, the proxy-variable for the type of school population (stimulation or non-stimulation school) did not explain any of the between-school variance, if controlled for individual differences in attitudinal and behaviourial aspects.

Table 3: Results of the regression analyses (* Not significant at p < .01)

| | Booklet | | |
Independent variable	A	B	C
intercept	40.15 (.27)	41.11 (.56)	28.06 (.53)
reading time	1.71 (.29)	.62 (.15)	.93 (.21)
comics	.11 (.05)		
reading process			.19 (.06)
reading preference			
reading behaviour			.55 (.23)
TV-time	.02 (.00)	.01 (.00)	
stimulation school			
gender			
language at home		1.11 (.42)	
only for low achievers			
reading time x achievement	-3.65 (.59)	-1.87 (.30)	-2.37 (.43)
TV-time x achievement	-.10 (.01)	-.08 (.01)	-.08 (.01)
R^2	.52	.60	.51
$R^2_{low\ achievers}$	.04	.08	.02*
$R^2_{high\ achievers}$	.04	.05	.03
$\sigma^2_{schools}$	2.84 (1.35)	.84 (.40)	.06 (.63)
$\sigma^2_{students}$	30.19 (2.08)	9.78 (.66)	18.83 (1.27)

The largest effect is for reading time in booklet A which equals 1.7. Per quarter of an hour reading each day the (expected) score raises by 1.7 points. So, in general, one has to read about three quarters of an hour to improve the score by one standard deviation. All other main effects are definitely smaller.A second point worth noting are the differences in main effects between booklets. Only the above-mentioned reading time has a significant contribution in all three booklets. There seem to be large differences between the three booklets in the way the achievement scores are related to attitude and behaviour scores. For instance, there is only a relation between the preference of books over comics and achievement in booklet A, and the effect t-time is absent in Booklet C. Apparently, different booklets tested different skills.

A third point we would like to mention concerns the regression weights for both interaction variables (reading time x achievement and t-time x achievement). Both effects are restricted to low achievers. They are relatively large and definitely differ from zero. Therefore, there has to be a non-linear relation between both aspects and reading achievement. While reading time and t-time seem to have a positive effect on achievement in the population, they have a relatively strong negative influence in the subpopulation of low achievers. Hence, both variables have a different meaning in the distinguished subpopulations. To visualise the results we have plotted the observed scores against the predicted scores for one of the booklets. This plot is presented in Figure 1.

What was inferred from Table 3 can be seen in Figure 1: in the subpopulation of weak readers the relation between both time variables and achievement differs from the same relations in the population as a whole. This results in regression lines with different slopes in both subpopulations. Two examples. Low achievers in booklet A who read an average half an hour each day are expected to score 7.3[1] points lower than high achievers who report they read the same amount of time

---

[1] The unit of measurement for reading time is a quarter of an hour. The expected score for a low achiever if one only looks at the reading time therefore equals $(1.71x2 - 3.65x2) = -3.88$, whereas the expected score for high achievers equals $(2x1.71) = 3.42$. The difference is 7.3 points.

every day, which is about 1.3 standard deviation. The second example concerns TV-time. The mean time spent watching television is about two hours. In booklet A, a low achiever who watches television for two hours a day is expected to score twelve points lower than a high achiever who watches for the same amount of time. This difference equals 2.1 standard deviation.

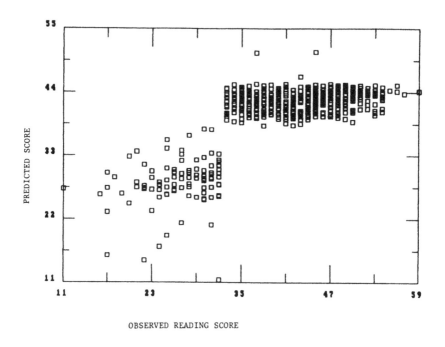

Figure 1: A plot of the estimated and the observed reading scores for booklet A

As might be expected the percentage of explained variance in the total sample is rather high. It varies around 50% across booklets. But it remains low, of course, in the two subsamples, as a consequence of the reduction of the variance in the dependent variable. For the low achievers who were assigned to booklet C, the percentage of explained variance does not significantly differ from zero at $p < .01$. For the other subpopulations R^2 varies from .04 to .08, which is in concordance with a correlation between .23 and .39.

A final point we would like to make concerns the between-school variation. For none of the booklets differences between schools could be assessed in the final model. All three variance estimates are smaller than 1.96 times the respective standard errors. Therefore, controlled for individual differences in attitudinal aspects, there is no difference in achievements between schools.

Discussion

We may conclude that there are relations between reading attitudes, reading behaviour and reading achievement. But at the individual level the picture is not very clear. The results do not completely converge; different aspects seem to contribute to the explanation of reading achievement in different booklets. Because in different booklets the students performed different tasks, it seems plausible to look for an explanation in tasks performed. In general this might be an explanation for the inconclusiveness of the research findings in the field: the relationships between attitudes, behaviour and achievement seem to (partially) depend on aspects of the reading tasks. It seems to be an open door – which only has to be forced – that attitudes always depend on the specific reading text. This implies that one has to speak of the relationship between attitudes, behaviour and achievement in terms of a clear specification of the dependent variable.

With respect to the first research question we have to stress the absence of (significant) differences between schools. Apparently, schools do not influence the relationship between affective and behaviourial characteristics and reading achievement. Strong emphasis on reading promotional activities does not seem to pay off in terms of achievements.

With respect to the second research question, we conclude that the relationship between behaviourial aspects and reading achievement is not invariant over subpopulations. Hence, these relationships are not linear in the population at large. In all three booklets the relationship between reading time, television time and reading achievement proved to be non-linear. These relationships seem to depend on the level of achievement. Therefore, the meaning of reading time and TV-time depends on the level of proficiency. This might be explained by the type of books read and the type of television programmes watched. We suppose that high achievers and low achievers differ in the kind of books they read, as well as in the kinds of television programmes they watch. But, of course, this hypothesis needs an empirical foundation.

Summarising: the relationship between attitudes, behaviour and reading achievement seems to be a relationship at the individual level. This means that individual variables, like cultural background, the opportunity to read 'good' books, membership of a library, etcetera, might be important variables. Of course, these variables can not be seen apart from social economic background. Concluding, if one tries to reveal the relation between activities towards reading and reading achievement, it seems that one has to measure social economic background at the individual level, in stead of using a proxy-variable like stimulation school (or not) at school level. If one uses the approach presented in this paper one may overestimate the effect of attitude on reading.

References

Aitken, M., & Longford, N. (1986). Statistical modelling issues in school effectiveness studies. *Journal of the Royal Statistical Society, Series A, 149*, 1-43.

Bergh, H. van den, & Kuhlemeier, H. (1989). Peilingsonderzoek in het voortgezet onderwijs [National assessment in secondary education]. *Handboek voor de basisvorming (3)*. Deventer: Van Loghem Slaterus.

Bos, D.J., & Oostdam, R.J. (1985). *Voorstudie periodieke peiling van onderwijsniveau: Het onderwijsaanbod Nederlands in de zesde klas* [A feasibility study to a national assessment in primary education]. Amsterdam: SCO.

Bryk, A.S., & Raudenbush, S.W. (1981). Toward a more appropriate conceptualization of research on school effects: A three-level hierarchical linear model. *American Journal of Education, 13*, 16-30.

Brett, A. (1978). The influence of effective education on the cognitive performance of kindergarten children. *Child Study Journal, 8*, 156-173.

Cialdini, R.R., Petty, R.E., & Cacioppo, J.T. (1981). Attitude and attitude change. *Annual Review of Psychology, 32*, 357-404.

Condon, M.W.F. (1978). Consideration of affect in comprehension: The person belongs in reading. *Viewpoints in Teaching and Learning, 54*, 107-116.

Cronbach, L.J. (1979). *Research in classrooms and school: formulations of questions, designs and analysis.* Occasional paper: Stanford Evaluation Consortium.

Daly, J.A. (1978). Writing apprehension and writing competency. *Journal of Educational Research, 72*, 10-14.

Fishbein, M. (1977). Attitudes and the prediction of behaviour. In M. Fishbein (Ed.), *Readings in attitude theory and measurement.* New York: Wiley.

Hoogeveen, M., & Verkampen, M (1985). *Schrijfonderwijs in praktijk. Een verslag van een etnografisch onderzoek naar de invoering van thematisch cursorisch schrijfonderwijs op een basisschool* [Writing education in practice]. Enschede: SLO.

Kahn, S.B., & Weiss, J. (1973). The teaching of affective responses. In R.M.W. Travers (Ed.), *Second handbook of research on teaching.* Chicago: Rand McNally.

Leidse werkgroep voor moedertaaldidactiek (1982). *Moedertaalonderwijs in ontwikkeling* [Mother tongue education in development]. Muiderberg: Coutinho.

Lewis, J. (1980). The relationship between attitude toward reading and reading succes. *Educational and Psychological Measurement, 40*, 261-262.

Miller, M.D., & Daly, J.A. (1975). *The development of a measure of writing apprehension.* Chicago: Educational Testing Service (Tests in microfiche: 006 172).

Prosser, R., Rasbash, J., & Goldstein, H. (1990). *ML3: Users' guide.* London: University of London.

Rosenberg, M.J. & Hovland, C.I. (1960). Cognitive, affective and behavioral components of attitudes. In *Attitudes, organization and change.* New Haven: Yale University Press.

Rijlaarsdam, G., & Bergh, H. van den (1987). *The development of the speaking apprehension measure.* Paper presented at the Second International Oracy Convention. Norwich: School of Education, University of East Anglia (ED 285 236).

Ward, B., Mead, N.A., & Searls, D.T. (1983). The releationship of students academic achievement to television watching, leisure time reading and homework. Education commission of the states. Denver, CO (ED 236 349).

Wicker, A.W. (1969). Attitudes versus action: The reliability of verbal and overt behaviour responses to attitude objects. *Journal of Social Issues, 25*, 41-78.

Part III

Processing and Learning Strategies

10

Uncertainty in text comprehension: Clues and effects

Lucia Lumbelli
University of Trieste (Italy)

Abstract

A specific feature of text comprehension processing was focused upon: the reader's feeling of uncertainty with regard to her/his own text reconstruction. This feeling has been proposed as a significant variable in research on metacomprehension (Markman, 1977, 1979; Baker, 1985) and on cognitive integration in comprehension and memory (Bransford & Johnson, 1972; Graesser, 1981; Van Dijk & Kintsch, 1983; Abbott et al., 1985).

The prediction was that if the reader realises that her/his processing is inadequate, transformation and cancellation is more likely to take place in her/his long-term memory, and therefore learning through text will be less effective. A specific procedure was designed to check this hypothesis: (l) poor readers were used as subjects, who would more likely present inadequate text elaboration and blockage of the automatic pilot (Brown, 1975), and hence conscious processing; (2) a 'natural' text was chosen whose sole criterion was that it contained some passage very likely to produce comprehension difficulties and inadequacies; (3) in the initial session an on-line procedure was adopted, in which the subjects were asked to read and repeat short text passages; in this way, concurrent verbal protocols (Ericsson & Simon, 1980, 1984) were collected on comprehension processes; (4) more complete and accurate protocols were pursued solely by reflection responses focused on subjects' spontaneous protocols found to be ambiguous and incomplete; (5) in a session carried out after a two-week interval, the subjects were given a free recall task and were once again encouraged to fill in any gaps by means of reflection responses. The hypothesis was confirmed, since there was a significantly higher number of transformed constructions (forgetting) among the aware non-comprehenders than among the unaware non-comprehenders.

Introduction

In earlier exploratory research (Lumbelli, 1987), an independent variable was identified which refers to a local feature of the text comprehension process: the uncertainty felt by the reader while processing certain passages of a text.

It seemed useful to discover more about the influence of this variable on global text comprehension and recall, in order to shed more light on the influence of schemes on both comprehension and recall (Bransford & Johnson, 1972; Miller & Kintsch, 1980; Graesser, 1981; Kintsch, 1982; Van Dijk & Kintsch, 1983; Trabasso *et al.*, 1984; Persig & Kintsch, 1985; Abbott *et al.*, 1985). Another reason for studying this variable is the guarantee of ecological validity (Neisser, 1982, 1986) it seems to offer for a study that sets out to answer the following question: What happens in the memory of a reader who has read a text with an awareness of her/his own comprehension difficulties?

Incomprehension with uncertainty can be seen as a clue to the reader's perception of some inconsistency. The psychological status of this inconsistency presents one main difference with respect to those inconsistencies constructed artificially in research on metacomprehension (Markman, 1977, 1979; Markman & Gorin, 1981): while in these studies the inconsistency is *definitely present in the text*, but *may not* be perceived by the subject (Baker, 1985), in the case of incomprehension with uncertainty it is the existence of the *inconsistency for the subject* which is guaranteed, even if this does not correspond to an actual text inconsistency, i.e., even if the representation of the inconsistency is caused by errors in the comprehension process.

It is our assumption that with respect to the effects on recall, the variable represented by incomprehension with uncertainty is more important than the variable (in)adequacy of comprehension. In fact, if the inadequate decoder lacks uncertainty, it means s/he has not noticed the coherence problem, and therefore *the situation for the subject will be identical*, both in this case and in the one in which comprehension is adequate. The (in)adequateness of the comprehension only emerges from the comparative analysis of the subject's verbal report on the one hand, and the corresponding segment of text on the other, including the integrations implied by bridging inferences (Clark, 1977). But this distinction has *no psychological reality*, and therefore no psychologically significant influence on memory processes.

Instead, our prediction was that the presence/absence of uncertainty in comprehension would have a significant influence on recall data: if the reader is unaware of having misunderstood, or does not perceive any inconsistency, i.e., processes the text information incorrectly but automatically, her/his performance in recall will be like the correct text comprehension; if, on the other hand, the reader realises that her/his processing is inadequate and approaches these difficulties with the same kind of awareness used to solve a problem, her/his memory is more likely to seek out more satisfactory solutions by transforming the original, unsatisfactory product with the help of her/his previous knowledge. Thus, when there is awareness of uncertainty, reconstruction by super-imposing schemes is more likely to occur than when the reader is unaware of having misunderstood. In other words, in the case of uncertainty in the evaluation of one's own comprehension, i.e., elimination or transformation of some parts of the comprehension protocols, forgetting would be more likely to occur.

This phenomenon could be seen as a very special kind of positive forgetting (Bjork, 1970, 1978; Geiselman, 1974, 1977; Cesa Bianchi & Cioffi, 1987): the 'positive' aspect in this case derives from the fact that forgetting helps the subject give coherence to information that had remained incoherent; in other words, forgetting gives coherence to a cognitive situation which is inconsistent and which is perceived to be so.

Method

The methodological strategy used in the present investigation may be seen as alternative and/or complementary to the *incongruity paradigm* mentioned above (Markman, 1977, 1979; Markman & Gorin, 1981; Capelli & Markman, 1982; Baker, 1985), undeniably more orthodox from the point of view of traditional experimental method. Instead of trying to get round the problem of how to gain access to the subject's knowledge of the experience, a direct approach is made here by the use of verbal protocols produced by the subject during the very act of processing information. The main points of the discussion on verbal protocols (While, 1988) were taken into account in designing our methodology.

According to Ericsson and Simon (1980, 1984) the question of the reliability of verbal protocols is presented in a very different way, depending on whether the verbalization is *concurrent* with the process being investigated, or whether it is *retrospective*, i.e., if it takes place when the process has finished. In the first case, there are fewer distorting factors and they vary according to the type of the cognitive processes examined and to the method of probing used: whether the questions are open-ended or of the closed type (where the answer can be inferred without recourse to the process being investigated), and whether the interview context is favourable to the production of *spontaneous* protocols.

Our study follows those premises this way:

(1) The subjects' verbal protocols regarding their text elaboration processes were rendered *as concurrently as possible*, by asking them to repeat segments short enough to ensure that the process was still being heeded in the course of verbalisation.

(2) Interference from pre-structured stimuli was avoided by telling subjects to report everything that came into their heads while reading, including impressions of uncertainty and difficulties of elaboration;

(3) Since a purely observative method can run the risk of being over-dispersive, the text was analysed beforehand to identify *the critical passage in which comprehension difficulties, and hence chances of uncertainty,*

were most likely to arise; i.e., that passage had: (3.1) to contain gaps to be completed by the reader through her/his previous knowledge and/or inferences from the context of the story, and (3.2) to allow the reader only to repeat a few bits of it, without having to choose between a complete repetition and a complete silence;

(4) The chances that comprehension difficulties would emerge in the subjects examined were increased by using poor readers for the experiment.

This latter choice is of the utmost importance, because all the stratagems found out to avoid interfering with the subject's verbal response will be useless if the process which the response is supposed to reflect is defined as highly automatic and hence extraneous to the awareness of the subject her/himself. This is one of the fundamental arguments against using verbal protocols to study the process of verbal comprehension. We take this into account by choosing those subjects who will most likely happen to be conscious of this process, because they will most likely face difficulties and therefore their so-called *automatic pilot* (Brown, 1975, 1978) will most likely stop functioning. Thus, using *poor* readers as subjects provides the experimenter with the richest possible source of information regarding the processes of verbal comprehension.

Finally, our integration consists of pursuing the best *completeness* and *accuracy* of the verbal reports by applying a form of interview defined and used in other research fields, the *interviewee-centred interview* (Rogers, 1945, 1951; Rogers & Kinget, 1967; Kahn & Cannel, 1957) and adapting it for our purposes. The most useful feature of this kind of interview is the *reflection response*, a form of utterance consisting of a *paraphrase of the subject's spontaneous protocol* (or of that part of it that is important and relevant to the purposes of the study) introduced by phrases of doubt. As a whole, it can be considered as *an indirect speech act of request for clarification and completion* (Lumbelli, 1987, 1989, 1990).

Subjects

27 subjects were examined, of approximately 13 years of age. They had all the following requisites: considerable reading difficulties despite the absence of any kind of cognitive deficiency; socio-cultural environment low enough to attribute them with a certain amount of verbal disadvantage; willingness to comply with the request to Think Out Loud (Olson, Duffy & Mack, 1984), and thus motivation to express themselves verbally and to engage in some cognitive activity for a relatively long period (the average length of each session was over one hour).

Each subject participated in two sessions. During the first session the text was presented in successive fragments. At every interval, the reader was invited to repeat what s/he had understood, adding any impressions of uncertainty and was then asked for integration by means of reflection responses. The initial instruction was:

> Now, we're going to read a story together. We're going to read it in segments. After every segment, we'll stop, and you'll tell me in your own words what you've understood about the piece you've just read; what has stuck in your mind, and also what you don't think you've understood, what you're not sure about. The important thing is that you say everything that comes into your head, even if it seems wrong or silly. Above all, it's important that you tell me when you feel uncertain about something. This isn't a test, where you have to give the right answer and avoid making mistakes, and where you usually only speak when you're sure you've understood correctly. Here, your doubts and uncertainties are very important.

At the end of every fragment the subjects were also given a *request for self-evaluation with a multiple-choice question*. They were asked: How sure are you that you have really understood this bit? Choose from the following answers: *I'm fairly sure, I'm very sure, I'm not very sure.*

We considered as clue of uncertainty in incomprehension protocols the choice of the answer *not very sure* to the final question.

The verbal protocols, both spontaneously provided and elicited by experimenter's reflection responses, were analysed. Almost every subject who had chosen the answer *not very sure* also provided utterances of uncertainty during the interview as well.

The second session took place two weeks after the first one. In this second session every subject was invited to recall the content and was also encouraged to complete it by reflection responses. This ensured that the subject was given the opportunity to look for the information in her/his mind and to retrieve information that was initially unable to emerge. The risk of cuing the answers was also avoided. In fact, with the reflection response, the proposals to continue exploring referred only to what the subject had already recalled.

Material

The subjects were presented with one of the short stories from Italo Calvino's Marcovaldo, entitled Father Christmas' Children. As regards the first part of the text, the following short summary has to be sufficient:

> Through a series of complex circumstances a rather strange thing had happened. Marcovaldo's children had taken as presents, to a very rich child, various objects they had managed to find in their poor home and which were undeniably very common: a hammer, a catapult and a box of matches. The rich child, who had been indifferent to the large number of other presents he had already received, had shown enormous enthusiasm for these strange and unusual presents and had used them at once to destroy everything he could find in the house.

The next piece of information is directly and clearly connected to this:

> Marcovaldo is desperate. Since the child's father is very important to the company he works for, Marcovaldo is almost sure he will lose his job because of the part his children played in the destruction of the rich child's house. Marcovaldo's fears are confirmed when he arrives at work and three office managers come up to him, one shouting: "Stop! Unload everything, immediately!"

This is where the final crucial segment comes, which has to provide the observative data for our investigation:

> "Quick! the parcels have to be changed!", said the managers. "The Union for increasing Christmas Sales has started a Campaign to launch the Destructive Gift!"
> "Just like that, out of the blue", one of them remarked. "They could have thought about it sooner."
> "It was an unexpected discovery by the president", explained another. "It seems that his child received some super-modern gift items, Japanese, I think, and for the first time they saw him having fun."
> "What counts above all", added the third, "is that the Destructive Gift can be used to destroy every kind of article: that's what we need to speed up consump-

tion and put some vigour back into the market ... All in no time at all and within the reach of a child ... The president of the Union sees new horizons opening up, he's beside himself with enthusiasm."

"But this child", asked Marcovaldo, in a whisper, "did he really destroy a lot of stuff?"

"It's difficult even to make a rough estimate, seeing that the house is still on fire."

The comprehension difficulties that resulted as the most probable from the preliminary text analysis are:

(1) difficulty to refer the phrase Destructive Gift to the peculiar presents Marcovaldo's sons had taken to the rich child (inferences from the fairly distant parts of the text);

(2) difficulty to refer the Japanese toys to those peculiar presents (inferences from the fact that the rich child had enjoyed both kinds of presents);

(3) difficulty to identify some *bridge* that guarantees coherence between the disruptive feature of the present to be launched and "the vigour back into the market" predicted as a consequence of his launch (inferences from knowledge about the connections among supply, demand and profit in the field of economics).

This part of the text plays a central role in the narrative structure of Calvino's story, because it informs the reader about the final result of the plot that begins with the decision of Marcovaldo's children to take presents to a poor child and continues with the series of misunderstandings that this decision triggers off. Anyone who has not understood this textual passage, has not understood the story. The corresponding macropropositions consist of the two reasons underlying the decision to market Marcovaldo's children's presents as the Destructive Gift: first, those presents had greatly amused the bored rich child; secondly, the idea emerged of using them for their destructive effect on other household objects, in order to increase sales of the latter. More fun for children and maximum potential for destroying many everyday consumer goods are the two reasons why there is a rush to market the unusual presents of Marcovaldo's children.

Results

In the first session 6 subjects produced protocols classified as clues of comprehension, while among those who failed to understand, 9 subjects were excluded from the data pertaining to the hypothesis, because they had given the response *fairly sure* after their attempts to reconstruct the meaning of the critical passage.

Thus, the data pertaining to the hypothesis being tested regarded 12 subjects, all of whom had shown they had not understood, and had given *not very sure* or *very sure* responses.

Our prediction was that the subjects who had responded *not very sure* would reveal more instances of forgetting, i.e., of transformation of the initial version. The subjects who had responded *very sure*, on the other hand, though they had not understood either, would reveal significantly fewer instances of forgetting. The results confirmed our prediction.

Table 1: Relationship between uncertainty in incomprehension and forgetting in poor readers

	Remembering (Information reconstruction maintained)	Forgetting (Information reconstruction changed)	Total
Incomprehension with certainty	4	0	4
Incomprehension with uncertainty	1	7	8
Total	5	7	12

Table 1 shows that, while the 4 subjects who had replied *very sure* retained the version given in the initial session without making changes, only 1 of the 8 *not very sure* subjects did so. Therefore, awareness of difficulty in understanding emerged as a significant factor determining a certain amount of memory transformations (Fisher Test: $P = 0.0101$).

To illustrate the quality of these results, the verbal protocols collected from two subjects are shown below, in the step-by-step reading session (first column) and then in the recall session (second column). The first subject had said that he was *not very sure* at the end of the reconstruction of the crucial passage of the text, and in the second session he produced data classifiable as forgetting, since they presented cancellations and transformations with respect to the comprehension protocols.

We have italicised those segments of both columns which consist of (a) text phrases faithfully repeated and incorrect integration carried out in the comprehension session and then not retrieved in the recall session; (b) new integrations emerging only in the recall session:

S. These three men go up to Marcovaldo and tell him to unload everything and Marcovaldo thought to himself that he had been sacked ... but not because *they had to change the parcels and take a destructive parcel* ... because they had heard that a little boy had received some Japanese electronic games like that and so it was the first time that he had been seen to ... that he had enjoyed himself ... so they wanted to change all the parcels ... and this other person ... the third person tells him they have to give presents but *in no time at all* and they *have to be within the reach of a child* and *put some vigour back into the market* ... then they talked about this child who had set fire to everything ... and so they were trying to redo all the parcels as quickly as possible so that this would give them a bigger advantage than the ones they had prepared before ... that is *the children would have more fun with these electronic games so different from the ones that little boy had received* ... after towards the end he says that *another little boy had set fire to everything.*

S. After I remember that they had got a bit angry because they had found out from this little boy's father that he had received presents and destroyed ... well not exactly destroyed but broken some things in the house ... and then I remember that they had to revolutionise the presents because they could see that they weren't very happy with the presents this father Christmas delivered ... so they wanted to change them and put other ones that would make people I mean the children happier

I. You said that at the company they got angry because they know this child did some damage to the house

S. I'm not exactly sure about that ... yes they're angry and *they want to find whoever gave the child these presents* that weren't exactly toys ... so they wanted to find out who.

I. And then you said they wanted to change the presents Marcovaldo was taking around and put presents in the packages which would make children happier

S. Yes I don't remember that part well because I got to the end of the story with what I said but before there were other things I can't remember ... yes in the end they decide *to change the presents because*

they see that this child had more fun and
they wanted to change the presents to be a
bit more original ... not give the usual
presents.

The second one had said he was *very sure* about his own reconstruction of the crucial textual passage (which was incorrect from the textual point of view) and in the second session he produced a summary of the content of the initial protocols without important lapses or changes. Here, we have italicised the integration in the comprehension session which was faithfully retrieved in the recall session:

A. Three bosses arrive and say to Marcovaldo: unload the parcels immediately because a new invention has been created ... the commendatore has decided on the destructive toy because his son received those presents and he didn't know that Marcovaldo's children had given him those toys and Marcovaldo felt a bit sacked ... so the commendatore said that when his son got those presents he had a lot of fun ... this destructive present was supposed to destroy all the other kinds of articles ... then the commendatore says to Marcovaldo: my son destroyed a lot of stuff and Marcovaldo says to him: your son destroyed a lot of stuff? And then the commendatore says to him: it's difficult to calculate because the house is al- most completely burnt down ... *inside himself the commendatore is not pleased that the house has been set on fire ... he's happy for his son because he hadn't seen him smile for a long time but not for the house set on fire.*

A. Marcovaldo goes into the company and three bosses arrive and he thinks they are there to sack him and then they say ... they say to him ... the commendatore tells him he has made this plan about the destructive toy because his son had set fire to everything and he looked happy for the first time ... *so the commendatore says to Marcovaldo now my son is happy but I've got all my house burnt down ...* that's the end of the story.

Discussion

Our prediction that uncertainty in the initial elaboration of the information would influence the corresponding free recall data was confirmed. The evidence about metacomprehension and cognitive integration in comprehension and memory mentioned above serves to explain this relation between uncertainty in comprehension and likelihood of memory transformation. If the reader expresses uncertainty, s/he is probably aware of the problem of coherence created by the inadequate elaboration or incomprehension. This reader shows more reading ability than the unaware non-comprehender because s/he knows how to apply the *internal consistency standard* to her/his own decoding processes, and is therefore in a more favourable position from the point of view of learning through texts. In fact, an awareness of inadequacy during the first information processing makes the subject more ready to ask and receive instructional help. This awareness of inadequacy is an important advantage only if this instructional opportunity exists. Our study shows what happens when there is inadequacy awareness without any instructional aid being programmed, and the poor reader is thus left alone with the text: the initial state of uncertainty prevents unambiguous and stable memory storage, and makes further influence of schemata and scripts more likely, as the experimental evidence shows. The results of the comprehension process are less stable and less resistant with respect to the action of the previous knowledge. Therefore, comprehension inadequacies with awareness have a snowball effect on whatever is left impressed on one's mind unless speedy and appropriate assistance is provided.

In the case of unaware incomprehension, there is an additional instructional problem. The lapses and transformations occur automatically and are thus smooth and immediate. This may be explained by the experimental evidence showing an inverse correlation between the strength of the *external consistency standard* and the level of reading ability. The lack of uncertainty about one's own inadequate comprehension processes is both a clue of reading poverty and a condition favourable to faithful retrieval of information, i.e., to greater stability in the initial incorrect reconstruction of the critical passage: everything that needed doing in order to guarantee coherence has already been done through automatic decoding in the initial elaboration of the information. No help is requested and the 'error' is stored permanently.

The educational suggestions that may be derived from this investigation are:

(1) To avoid the snowball effect on the memory, it is necessary to provide the requested correction and explanation during the *initial elaboration of the information, before the long-term memory processes are triggered*;

(2) If the initial incomprehension is unconscious, *preliminary stimuli should be provided* to bring about a consciousness of one's own inadequacy, i.e., the reader should be made to notice the inconsistency her/his cognitive integration has caused or has failed to eliminate.

In conclusion, we should like to refer briefly to a methodological issue which deserves much wider discussion. Our results are statistically significant but the number of subjects is very low. This is the price that has to be paid for the benefit of ecological validity. The length and complexity of the data-gathering procedure makes it difficult to extend it to a larger number of subjects. On the other hand, there is the additional benefit of solving the problem still posed by the *multistandard method* in research on metacomprehension: how to make sure that the experimenter does not bias the subjects' verbal protocols through the instructions or the task proposed, and at the same time avoid an excessive dispersion of data. The solution lies in deciding to focus on a single, clearly-defined text passage which is 'naturally' difficult to comprehend, and to use only the subjects' protocols related to that specific passage as experimental data.

References

Abbott, V., Black, J.B., & Smith, E.E. (1985). The representation of scripts in memory. *Journal of Memory and Language, 24,* 179-199.

Baker, L. (1985). How do we know when we don't understand? Standards for evaluating text comprehension. In D.L. Forrest-Pressley, G.E. Mackinnon & T.G. Waller (Eds.), *Metacognition, Cognition, and Human Performance,* vol. I (pp. 155-205). Orlando: Academic Press.

Bjork, R. (1970). Positive forgetting: The non-interference of items intentionally forgotten. *Journal of Verbal Learning and Verbal Behavior, 9,* 255-268.

Bjork, R. (1978). Theoretical implications of directed forgetting. In A. Melton & E. Martin (Eds.), *Coding Processes in Human Memory.* New York: Wiley.

Bransford, J.D., & Johnson, M.K. (1972). Contextual prerequisites for understanding: some investigations on comprehension and recall. *Journal of Verbal Learning and Verbal Behavior, 11*, 717-726.

Brown, A.L. (1975). The development of memory: Knowing about knowing, and knowing how to know. In H.W. Reese (Ed.), *Advances in Child Development*, vol. 10. New York: Academic Press.

Brown, A.L. (1978). Knowing when, where and how to remember: A problem of metacognition. In R. Glaser (Ed.), *Advances in Instructional Psychology*, vol. I. Hillsdale, NJ: Lawrence Erlbaum Associates.

Capelli, C.A., & Markman, E.M. (1982). Suggestions for training comprehension monitoring. *Topics in Learning and Learning Disabilities, 2*, 87-96.

Cesa Bianchi, M., & Cioffi, G. (1987). Il concetto di oblio e di dimenticanza volontaria. *IKON, 15*, 25-88.

Clark, H.H. (1977). Inferences in comprehension. In D. Laberge & S.J. Samuels (Eds.), *Basic Processes in Reading: Perception and Comprehension*. Hillsdale, NJ: Lawrence Erlbaum Associates.

Ericsson, E.A., & Simon, H.A. (1980). Verbal protocols as data. *Psychological Review, 87*, 215-251.

Ericsson, K.A., & Simon, H.A. (1984). *Protocol Analysis*. Cambridge, MA: MIT Press.

Geiselman, R. (1974). Positive forgetting of sentence material. *Memory and Cognition, 2*, 677-682.

Geiselman, R. (1977). Effects of sentence ordering on thematic decisions to remember and to forget prose. *Memory and Cognition, 5*, 323-330.

Graesser, A.C. (1981). *Prose Comprehension beyond the Word*. New York: Springer.

Kahn, R.L., & Cannel, Ch.F. (1957). *The Dynamics of Interviewing*. New York: Wiley.

Kintsch, W. (1982). Memory for texts. In A. Flammer & W. Kintsch (Eds.), *Text Processing*. Amsterdam: North Holland.

Lumbelli, L. (1987). Dalla difficoltà di capire alla dimenticanza volontaria. Una ricerca esplorativa. *IKON, 15*, 89-124.

Lumbelli, L. (1989). *Fenomenologia dello scrivere chiaro*. Roma: Editori Riuniti.

Lumbelli, L. (1990). Controllo ed autocontrollo nella comprensione verbale. *Orientamenti pedagogici, 3*, 512-523.

Markman, E.M. (1977). Realizing that you don't understand: A preliminary investigation. *Child Development, 48*, 986-992.

Markman, E.M. (1979). Realizing that you don't understand: Elementary school children's awareness of inconsistencies. *Child Development, 50*, 643-655.

Miller, J.R., & Kintsch, W. (1980). Readability and recall of short prose passages. *Journal of Experimental Psychology. Human Learning and Memory, 6*, 335-354.

Neisser, U. (1982). *Memory Observed*. San Francisco: Freeman.

Neisser, U. (1986). Nested structure in autobiographical memory. In D.C. Rubin (Ed.), *Autobiographical Memory*. New York: Cambridge University Press.

Olson, G.M., Duffy, S.A., & Mack, R.L. (1984). Thinking Out Loud as a method for studying real-time comprehension processes. In D.E. Kieras & M.A. Just (Eds.), *New Methods in Reading Comprehension Research*. Hillsdale, NJ: Lawrence Erlbaum Associates.

Perrig, W., & Kintsch, W. (1985). Propositional and situational representation of text. *Journal of Memory and Language, 24*, 503-518.

Rogers, C.R. (1945). The non-directive method as a technique in social research. *American Journal of Sociology, 50*, 279-283.

Rogers, C.R. (1951). *Client-Centered Therapy*. New York: Houghton Mifflin.

Rogers, C.R., & Kinget, M. (1967). *Psychothérapie et relations humaines*. Louvain: Nauwelaerts.

Trabasso, T., Secco, T., & Van den Broek, P. (1984). Causal cohesion and story coherence. In H. Mandl, N.L. Stein & T. Trabasso (Eds.), *Learning and comprehension of text*. Hillsdale, NJ: Lawrence Erlbaum Associates.

White, P.A. (1988). Knowing more about what we can tell: 'Introspective access' and causal report accuracy 10 years later. *British Journal of Psychology*, *79*, 13-45.

11

The development and the usage of some particular cognitive strategies concerning text processing in physics and in history

Anna Aventissian-Pagoropoulou

University of Athens (Greece)

Abstract

The ability of ninth-graders to take advantage of task-tailored cognitive strategies was investigated. Two types of expository text were provided, which were assumed to be quite demanding for the students to cope with. Readers at the age of fifteen were not expected to be strategic, especially those of the medium and the low academic levels. The intervention method that was proposed to the experimental groups in the experiment was examined as to its effects. The impact of the intervention method was highlighted regarding the two types of expository texts, the academic ability levels of the students, and the two different conditions (i.e., co-operative vs. individual) under which the processing of the texts took place.

Introduction

In considering some real-world tasks of learning from text, we will find a variety of specific applications in diverse domains such as language, mathematics, science, social studies, etc. (Pressley, Levin & Bryant, 1983). Data from different studies offer support for the notion that the ability to cope with texts is developmentally sensitive, with clear improvement occurring during the transition from childhood to adulthood. During the adolescent years, there is an increased sophistication in the use of elaborative strategies, which is predictive of recall performance (Brown & Smiley, 1977).

On the other hand, the processing of the expository texts seems to be quite a demanding task for high school students. Most of the high school teachers from many different countries complain that their students cannot cope with their textbooks, in order to be able to comprehend sufficiently and memorise a text passage for future use. But the students also complain that the technical terms contained in some texts (e.g., Physics, Chemistry or Biology), or the arguments and the reasoning contained in other texts (as, for instance, in History or Sociology) often put them in a difficult position and make their study quite an unpleasant task (Meyer, Brandt & Blunt, 1980; Chi, Feltovich & Glaser, 1981).

Hence, the interest of cognitive psychologists has arisen who, in turn, tried to find out what may be wrong with the students, or the texts, or both and, as a result, why the interaction between readers and texts is so poor. Moreover, many cognitive psychologists in recent decades have tried to intervene in this interaction and help higher-school students with the difficulties they are confronted with when processing expository texts.

Reading stages and cognitive strategies after middle childhood

Reading ability proceeds from a kind of pseudo-reading to the stages of "learning to read" and, subsequently, the stages of "reading (in order) to learn", to the final stage when reading is a highly creative undertaking. Students generally progress in characteristic ways from stage to stage, as changes occur with age and educational experience (Chall, 1983). Up to the age of fourteen, students are expected to have grown in their ability to analyse what they read and to react critically to the different viewpoints they come across.

Within this age level, some common characteristics are to be found pertaining to the cognitive strategy component. Despite the fact that evidence bearing on strategy development during adolescence is relatively sparse, the findings that have emerged suggest some interesting progressions in memory strategy usage during adolescence. More specifically, it is found that *some* adolescents spontaneously employ strategies that younger children do not, and despite the pronounced individual differences, those adolescents who do not use sophisticated strategies spontaneously will often benefit from explicit instruction in their use (Pressley, Levin & Bryant, 1983). Based on the above, it is useful to see what happens when some particular cognitive strategies are provided to adolescents, together with explicit oral instruction on how to use them.

The aim of the study

The aim of this study was to activate students while they were processing the texts by means of a cognitive strategy activator. An active student is supposed (Gagné, 1985) to be an active processor of the information, and this, in turn, is expected to yield some side effects pertaining to the student's: (a) activation of prior knowledge; (b) modification of his/her particular cognitive style, which tends to be more reflective, field-independent and attention-focusing; and (c) orientation towards the effort to attach the new information to his/her pre-existing knowledge.

In this way, the student will not rely heavily on either the new information coming from the text, nor on his/her pre-existing knowledge. The cognitive strategy activator in this study had the *form* of (a) a cloze procedure, so that the student was invoked to fill in gaps while processing the text; and (b) an advanced organiser, so that the hints given to the students were of a "higher" level than the propositions included in the texts themselves.

As to its *function*, the cognitive strategy activator was supposed to (a) clarify the content of the text passage (content-clarifying strategy); and (b) sensitise the subjects as to the more refined concepts in the texts, and the concepts' implicit meaning (Levin & Pressley, 1981).

By means of the activator described, the subjects were expected to process the texts at a deeper level and gain useful insights into it.

Research design

The research design included five variables (A, B, C, D, E) which represented:

- A the academic ability of the students as reported by their school teachers, in three levels (high, medium, low)
- B the cognitive strategy activator in two levels (presence or absence of the activator)
- C the mode of studying, namely the co-operative processing versus the individual processing in two levels (co-operation, non-co-operation)
- D the text material, namely the science text versus social science text, in two levels (Physics, History).
- E the score obtained on a questionnaire, consisting of probe questions (viz. "wh" questions).

Of the above variables, two were independent and fixed (A and D), two were independent and random (B and C), and one was dependent (E).

Method

Experimental texts

The two experimental texts were constructed according to the following criteria: (a) they matched the student's current cognitive status; and (b) they contained concepts that were quite familiar to the average student, while the text itself was novel.

For these conditions to be covered, the text in Physics was selected to be "The steam engine", and the text in History was selected to be "The British Empire in the 19th century".

The ninth-grade subjects in Greece are supposed to be familiar with concepts like force, moment, chemical energy, thermal energy, and the like. Moreover, the steam engine itself is described in their textbook, as it is applied to transport (e.g., trains). The correspondent experimental material was describing a primary type of steam engine and its use to pump water from mines.

The same principle holds true for the History text. The ninth-graders had already been taught in previous years about colonialism in their history of Ancient Greece, and were quite familiar with the topic. However, the colonialism in recent centuries and the establishment of the British Empire was a totally unfamiliar topic to them. In this way, we made sure that all of our subjects started with the text processing from a common basis.

Intervention texts / Strategies

The intervention material, i.e., the cognitive strategy activator, can be described as a mixture of text and diagram. This material was made to provide additional information about the text, such as principles, causal relations, assumptions, inferences etc. Moreover, the intervention material was provided in two phases in the experiment, by means of pamphlet No 1 and pamphlet No 2.

The first pamphlet (No 1) was given to the subjects of the experimental groups together with the text to be processed. There were only hints given in the first pamphlet, for instance, the subjects would read the beginning of the following sentence: "According to the physics principle of energy conversion, the chemical energy is converted into ....". The subjects were expected to complete the sentence, while, in parallel, they were processing the text.

In the second phase of the experiment, pamphlet No 2 was given to the subjects. This was the full version of the activator, that is, no gaps were contained in it. During the second phase, the subjects were asked to correct themselves and/or fill in the gaps in pamphlet No 1, i.e., to act in the way they had not been able to during the first phase of the experiment. For this latter task to be executed they were provided with a pen of a different colour than the one they had been using before.

In this way, each subject was providing a clear image of the processing he was able to do by himself and with the aid of our signals (hints), and the kind of processing he was not able to do. For those segments of the text that a subject was not able to process adequately, there was the full version of the activator, providing explicit information and help.

Intervention in the control groups

The control groups of the study received some kind of intervention, too. In this case, the intervention consisted of oral instruction; the subjects were asked to process the text in the way they usually did. They were instructed to take a chance and apply any kind of strategy that they were used to already. They were encouraged, for instance, to take notes, or underline, or write comments in the margins, or represent mentally the content of the text in the way of their choice.

The major advantage of this intervention was that the subjects were given the opportunity to use spontaneously the strategy of their choice, which they knew by experience to work.

Subjects

A total of 288 subjects participated in the experiment. Half of the sample, i.e., 144 subjects, processed the Physics text and the other half the History text. Again, half of them, i.e. 72 subjects, processed it co-operatively and the other half processed it individually. All three levels of the academic ability variable were represented in the groups. The control and the experimental groups were established according to the absence or presence of the cognitive strategy activator.

Procedure

The participants of the study were evaluated academically by their teachers. Next, the different groups of the experiment were established according to the participants' academic ability level in the two subjects (Physics/History). All three academic levels were present in the groups consisting of six students. The experiments took place in the libraries of the schools, where the following arrangement had been set up: For the cooperative groups, six chairs were placed around a table; after the processing procedure was over,the students were seated apart from one another, in order to answer the questionnaire. For the groups of individuals, the participants were separated from the very beginning.

After the whole procedure was over, the students were given an opportunity to talk with the experimenters about their experiences.

Scoring

Four different evaluators were involved in the scoring of the two questionnaires: two of them were experts in the field of Physics, correcting the protocols in Physics, and the other two were experts in the field of History, correcting the protocols in History. The questionnaires consisted of 20 test questions each. For each one of the questions a grade was given on a four-point scale (0-3); thus, the total sum of grades that a question could receive by the two evaluators ranged from 0 to 6.

Both qualitative and quantitative indexes were used as criteria for scoring. The grade of agreement between the two evaluators in the indexes was 90%, usually arriving at a consensus in the differences found.

Results and conclusions

There were three research hypotheses in the present study, pertaining to the impact of the cognitive strategy component in conjunction with: (1) the specific type of the expository texts; (2) the differing academic ability of the students participating in the task; and (3) the alternative condition under which text processing took place.

The results of the study stem from the four-way analysis of variance conducted (see Table 1), and the multiple comparison tests. Our factorial experiment had the form of 3×2×2×2, therefore a four-way ANOVA was conducted, involving the factors A, B, C and D. The significant F-values obtained in Table 1 indicated that all variables had an effect on students' text processing. However, multiple comparisons between parts of group means were performed to determine the particular groups which differed from each other.

The conclusions drawn from the above analyses of the data can be summarised as follows: (1) Subjects of medium academic ability benefited most by making the most of the strategy employed in the study, as regarding both types of expository text (see Table 2). (2) Subjects processing the Physics text were found to be more systematically affected by the strategy proposed than subjects processing the History text. In other words, both high and medium academic levels differed from their controls, and this finding was supported in two cases (interaction effect among ABCD variables and ABD variables) (see Table 3). (3) Subjects who processed the texts co-operatively were found to perform better than subjects who processed either text individually. In the interaction effects, however, no sign of significant differences were found between the two groups; hence, little sign of significant improvement was achieved by means of the communicative learning activity (see Table 4).

Table 1: Analysis of variance (* p <.05; ** p <.01; *** p <.001)

Source	df	MS	F
Academic Ability	2	57715.526	295.804***
Strategy	1	5599.740	28.700***
Co-operation	1	531.180	2.722*
Text	1	114.835	.589
Ability × Strategy	2	4696.981	8.697**
Ability × Co-operation	2	91.019	.466
Ability × Text	2	753.333	3.861*
Strategy × Co-operation	1	79.253	.406
Strategy × Text	1	81.361	.417
Co-operation × Text	1	27.468	.141
Abil. × Str. × Co-op.	2	70.855	.363
Abil. × Str. × Text	2	97.344	.499
Abil. × Co-op × Text	2	43.728	.224
Str. × Co-op × Text	1	76.453	.392
Abil. × Str. × Text × Co-op	2	195.114	4.814**
Residual	263	625.041	
Total	286	587.393	

Table 2:   Interaction of factors A, B, C and D

| | | TEXT | | | |
| | | PHYSICS | | HISTORY | |
	Academic ability	X	SD	X	SD
Experimental group	HIGH	103.08	12.4	94.66	16.4
	MEDIAN	98.66	17.6	91.25	10.2
	LOW	47.66	6.9	50.16	18.4
Control group	HIGH	92.08	19.2	91.04	5.09
	MEDIAN	79.83	12.5	77.83	4.78
	LOW	46.05	12.7	49.58	13.48

Table 3:   Interaction of factors A, B and D

| | | STRATEGY ACTIVATOR | | | |
| | | EXPERIMENTAL GROUP | | CONTROL GROUP | |
	Academic ability	X	SD	X	SD
Co-operative	High Physics	103.08	12.4	92.08	19.2
	Median Physics	98.66	17.6	79.83	12.5
Individual	Median Physics	93.83	19.2	76.50	11.4
	Median History	90.16	10.9	67.25	7.03

Table 4:   Interaction of factors B and C

| | STRATEGY ACTIVATOR | |
| | EXPERIMENTAL GROUP | CONTROL GROUP |
	X	X
Co-operative	80.96	72.45
Individual	78.81	69.27

Discussion

A relatively large spectrum of relevant factors influencing text processing was considered in the present study. The outcomes of learning from texts were based on reader variables, text characteristics, and the process of text comprehension itself (Mandl & Schnotz, 1986). As for the readers, the domain-specific prior knowledge was taken into account, in relation to specific strategies (Glaser, 1982). The readers themselves were 15-year- old children, and the comprehension-fostering activity they were involved in had obvious positive effects. As Meyer (1975) described, readers have to recognise the higher-order structure of a text in order to succeed with text comprehension.

Moreover, the structural aids provided are only effective if the learner has the possibility to use them actively; in the present study, this was achieved by means of the two pamphlets, which activated the students and which put emphasis on certain content elements and semantic relations contained in the texts to make the structural properties of the texts more obvious (Mandl & Schnotz, 1986).

The cognitive strategy activator utilised in this study was more than a single learning strategy or technique. Studies conducted on individual techniques are not effective in promoting comprehension and memory performance (Mandl, 1981, 1984; Weinstein & Mayer, 1985). Our cognitive strategy activator was supposed to be a more complex learning strategy programme, in which both elaborative and reductive techniques were included. The advantage of the experimental groups in adapting this strategy program was made clear in the results.

On the other hand, the control groups were given the opportunity to use the preferable processing strategy each subject would choose according to the proposed text organisation (Schnotz, 1984).

Finally, an attempt was made to introduce socio-cultural aspects in text processing by means of co-operation. Subjects in one half of the sample were seated around a table while processing the text; each member of those co-operative groups was instructed to interact during text processing, i.e., to exchange ideas, to express his/her queries, to make statements about the text read, and so on. In this way, more viewpoints were supposed to be introduced by the members of the co-operative groups than by any single person. It was expected (Mandl & Schnotz, 1986) that the social component in learning would have the added advantage that more thorough text processing would occur in the co-operative groups.

As a matter of fact, the underlying differences could possibly be shown if the co-operative groups had been studying in the above setting and the individual groups had been examined in the usual setting, i.e., doing their study at home.

In the present study, however, the members of the individual groups were also seated in the school library, although separate from one another. The school library is supposed to be a social context, too.

Obviously, more research is needed in order to confirm specific hypotheses on text processing. Both in the field of cognitive strategy research and in the field of co-operative learning, more attempts should be made which in turn may prove to be very promising for better and more thorough text processing.

References

Ausubel, D.P. (1980). Schemata, cognitive structure, and advanced organizers: A reply to Anderson, Spiro, and Anderson. *American Educational Research Journal, 17*(3), 400-404.

Anderson, J.R. (1983). *The Architecture of Cognition*. Harvard University Press.

Anderson, T.H., & Armbruster, B.B. (1984). Studying. In P.D. Pearson (Ed.), *Handbook of Research on Reading* (3rd ed.). New York: Longman.

Atwell, M., & Rhodes, Lynn K. (1984). Strategy lessons as alternatives to skills lessons in reading. *Journal of reading, 27*.

Bereiter, C., & M. Scardamalia (1982). From conversation to composition: The role of instruction in a developmental process. In R. Glaser (Ed.), *Advances in Instructional Psychology (Vol. 2)*. Hillsdale, NJ: Lawrence Erlbaum Associates.

Brown, A.L., & Smiley, S. (1977). Rating the importance of structural units of prose passages: A problem of metacognitive development, *Child Development, 48*, 1-8.

Clewell, S., & Haidemos, J. (1983). Organizational strategies to increase comprehension. *Reading World, 22*, 314-21.

Chall, J.S. (1983). *Stages of Reading Development*. McGraw-Hill Book Company.

Chi, M., Feltovich, P.J., & Glaser, R. (1984). Categorization and representation of physics problems by experts and novices. *Cognitive Science*, 121-152.

Day, J. (1983). The zone of proximal development. In M. Pressley & J. Levin, *Cognitive Strategy Research, Psychological Foundations*. New York: Springer-Verlag.

Fisher, P., & Mandl, H. (1982). Metacognitive regulation of text processing: Aspects and problems. In Flammer & W. Kintsch (Eds.), *Discourse processing*. Amsterdam: North-Holland.

Finley, F. (1983). Students' recall from science text. *Journal of Research in Science Teaching, 20*, 247-59.

Friedlander, F. (1982). Alternative modes of inquiry. *Small Group Behavior, 13*, 428-40.

Friedrich, H.F., Fischer, P.M., Kramer, D., & Mandl, H. (1985). Development and evaluation of a program facilitating comprehension of text. In G. d'Ydewalle (Ed.), *Cognition, Information Processing, and Motivation*. Amsterdam: Elsevier.

Gagné, E.D. (1985). *The Cognitive Psychology of School Learning*. Little, Brown and Company.

Glaser, R. (Ed.) (1982). *Advances in Instructional Psychology (Vol. 2)*. Hillsdale, NJ: Lawrence Erlbaum Associates.

Griffin, P., & Cole, M. (1984). Current activity for the future. The zo-ped. In B. Rogoff & V. Wertsch (Eds.), *Children's Learning in the Zone of Proximal Development*. Jossey-Bass.

Hildyard, A., & Olson, D. (1982). On the structure and meaning of prose text. In Th. Anderson & B. Armbruster (Eds.), *Reading Expository Material*. Wiley & Sons.

Kirby, J. (1984). Educational goals of cognitive plans and strategies. In J. Kirby (Ed.), *Cognitive strategies and educational performance*. Academic Press.

Kraft, R. (1985). Group-inquiry turns passive students active. *College Teaching, 33*.

Larkin, J.H., McDermott, J., Simon, D., & Simon, H.A. (1980). Models of competence in solving physics problems. *Cognitive Science, 4*, 317-345.

Levin, J., & Pressley, M. (1981). Improving children's prose comprehension: Selected strategies that seem to succeed. In C. Santa & B. Hayes (Eds.), *Children's Prose Comprehension*. International Reading Association.

Levin, J.R. (1976). What have we learned about maximizing what children learn? In J.R. Levin & V.L. Allen (Eds.), *Cognitive Learning in Children: Theories and Strategies*. New York: Academic Press.

Mandl, H., & Schnotz, W. (1986). New directions in text comprehension. In E. De Corte, J.G.L.C. Lodewijks, R. Parmentier & P. Span (Eds.), *Learning and Instruction*. Oxford: Pergamon Press.

Meyer, B.J.F. (1975). *The Organization of Prose and Its Effects on Recall*. Amsterdam: North-Holland.

Meyer, B., Brandt, D., & Blunt, G. (1980). Use of top-level structure in text: Key for reading comprehension in ninth-grade students. *Reading Research Quarterly*.

Nicholson, T. (1984). Experts and novices: A study of reading in the high school classroom. *Reading Research Quarterly*.

Palincsar, A., & Brown, A. (1984). Reciprocal teaching of comprehension-fostering and comprehension-monitoring activities. *Cognition and Instruction*.

Paris, S., & Lipton, W. (1983). Becoming a strategic reader. *Contemporary Educational Psychology, 8*.

Pressley, M., Levin, J.R., & Bryant, S. (1983). Memory strategy instruction during adolescence: When is explicit instruction needed? In M. Pressley & J. Levin (Eds.), *Cognitive Strategy Research, Psychological Foundations*. New York: Springler-Verlag.

Raphael, T., & McKinney, J. (1983). An examination of eighth- and ninth-grade children's question-answering behavior: An instructional study in metacognition. *Journal of Reading Behavior, 15*, 67-86.

Richardson, J., & Jones, G. (1984). Learning about learning. *Assessment and Evaluation in Higher Education, 9*.

Rothkopf, E. (1972). Structural text features and the control of processes in learning from written material. In R. Freedle & J. Carrol (Eds.), *Language Comprehension and the Acquisition of Knowledge*. Winston.

Schnotz, W. (1985). Selectivity in drawing inferences. In G. Rickheit & H. Strohner (Eds.), *Inferences in text processing*. Amsterdam: Elsevier.

Spiro, R. (1977). Remembering information from text: The state of schema approach. In R.C. Anderson, R. Spiro & W. Montague (Eds.), *Schooling and the Acquisition of Knowledge*. Hillsdale, NJ: Lawrence Erlbaum Associates.

Scherr, F. (1982). *Comprehending procedural instructions: The influence of metacognitive strategies*. Paper presented at the Annual meeting of the American Educational Research Association.

Schnotz, W. (1984). Comparative instructional text organization. In H. Mandl, N.L. Stein & T. Trabasso (Eds.), *Learning and Comprehension of Text* (pp. 53-81). Hillsdale, NJ: Lawrence Erlbaum Associates.

Taylor, B., & Beach, R. (1984). The effects of text structure instruction on middle-grade students' comprehension and production of expository text. *Reading Research Quarterly, 19*.

Van Dijk, A., & Kintsch, W. (1983). *Strategies of discourse comprehension*. Academic Press.

Wilson, J. (1981). *Student learning in higher education*. Wiley & Sons.

12

Teaching active text processing strategies: Some experimental results.

Juan A. García Madruga, Jesus I. Martín Cordero,
Juan L. Luque and Carlos Santamaria
University of Distance Education, Madrid (Spain)

Abstract

Current theories about comprehending and memorising texts comment on the relevance of active processing during reading. This active processing is necessary to achieve a coherent representation that includes the construction of a macrostructure and a situational model of the text content. This chapter describes an instructional programme that is designed to improve two important strategies: (a) drawing out main ideas from texts, and (b) constructing outlines (structural summaries of texts).

Taking this theoretical perspective, an experiment was performed to evaluate the impact of a training programme on the strategies mentioned above of secondary school students. In a pretest session, subjects were randomly assigned to three experimental conditions: advance outline *(where subjects were given a previous outline of the text),* outline construction *(in which they were required to elaborate the outline by themselves), and* no outline *(the control condition). They were given a four-session instruction programme. The first two sessions included a brief explanation of what the main ideas of the text meant and practice in drawing them out from different texts. The two final sessions consisted of practice on the elaboration and use of text outlines based on the main ideas. Again, different texts were used for this purpose. Finally, subjects were given a posttest session to test the effects of training, in which the preliminary experimental conditions were reinstated.*

This article reports some of the main results from this experiment, as well as a discussion of their implications for an educational framework.

Introduction

Studies carried out during the past 25 years by cognitive psychologists in the field of discourse comprehension, especially text comprehension, have defined the constructive and interactive nature of this process. The final outcome of comprehension is the construction of a semantic representation that includes a mental situational model of the state of affairs described in the text, integrating the contents of the texts with the subject's prior knowledge (Carpenter & Just, 1987; Johnson-Laird, 1983; Kintsch, 1988; Van Dijk & Kintsch, 1983). This constructive nature of discourse comprehension had already been pointed out by Bartlett's pioneering studies and corroborated by Bransford and collaborators (Bransford & McCarrell, 1975) and by schema theory (Rumelhart, 1980). The point is that text comprehension is an exceedingly complex task in which many cognitive processes on diverse levels (word recognition and lexical access, syntactic and semantic analysis) must interact. This issue is computationally approachable and solvable, based on the subjects' prior knowledge. Therefore, in order to resolve the problem of interpreting the linguistic input consistently, the subject must activate his/her knowledge of the material being dealt with, employ his/her strategic knowledge and recognise the rhetorical structure of the text (Garner, 1987; Resnick, 1984).

In addition, text comprehension is the result of a long acquisition process in which the child learns to automatise the more superficial levels freeing cognitive resources that can be applied to the construction of a precisely detailed mental representation. During this developmental process there comes a time when the child evolves from "learning to read" to "reading to learn". At this point comprehension is not just an end in itself but becomes a means of learning new information. From this moment on, the adolescent is faced with the task of understanding expository texts that try to transmit new contents in a complex organisation that the reader is not yet familiar with. In order to construct an accurate representation of this type of expository text retaining its main ideas, the subject must have a set of strategic skills that enable him/her to construct an adequate macrostructure of the text; he/she must also adapt to its rhetorical structure and follow the text's guidelines. It is not surprising that many subjects faced with a task beyond their capabilities opt for a repetitive kind of learning that sooner or later leads them to academic failure. The conclusions that can be drawn from this panorama are obvious. If we want to overcome the problems that learning from texts pose to a significant number of subjects of all ages, we must improve either the texts themselves by making their structures clearer and introducing different types of aids that facilitate comprehension, or the strategies and abilities that are vital to the subjects' comprehension. As stated earlier (García Madruga & Martín Cordero, 1987; García Madruga & León Cascón, in press) this is not necessarily an either/or proposition. This paper attempts to make progress in the development of simultaneous intervention of both texts and subjects. Before addressing our study, however, let us take a look at some of the salient points of the research in both areas.

Intervention in the text and the subjects

Studies on text intervention, both of internal improvement (structure, syntactic and lexical complexity) and the introduction of aids (objectives, questions, summaries, advance organisers, signalling) follow a pattern in which the results of this improvements depends on the measures employed (free recall, recognition, problem solving, etc.) along with the subjects' knowledge and skills. Nevertheless, it appears to be possible to facilitate text comprehension by improving them and by introducing certain aids such as advance organisers, which are a representative cognitive technique. Advance organisers seem to be effective in complex texts whose structures are not particularly explicit, and with students lacking special knowledge of the topic and the necessary strategic skills (Ausubel, 1978; Barnes & Clawson, 1975; García Madruga & Martín

Cordero, 1987; Mayer, 1979, 1983). Mayer (1979) has suggested that the function of advance organisers might be to facilitate the availability and to activate the prior knowledge that allows the assimilation of new information presented in the text. However, as has recently been proposed, the efficiency of advance organisers depends on the use the subjects really make of them since simply presenting them could be insufficient (Kloster & Winne, 1989).

We have found similar results (García Madruga, Luque Vilaseca & Martín Cordero, 1989; Martín Cordero, García Madruga, Luque Vilaseca & Santamaría Moreno, 1991) when attempting to test the efficiency of two aids that are clearly different, due both to their functions and to the research tradition from which they stem. These are objectives and structural summaries or outlines. Outlines explicitly provide the macrostructure and the top-level rhetorical structure of the text, and have been shown to be more effective than objectives in structural recall and comprehension measures. They are similar to advance organisers in that they structurally organise the texts' main ideas. However, our earlier studies suggested areas for future research into the type of propositional analysis used, which was molar with rather large idea-units, and into the subjects' actual use of the aids provided. We have tried to address both of these aspects in this study.

In the past fifteen years, numerous studies on the improvement of subjects' strategies and knowledge of the rhetorical structures of expository texts have shown that, despite the difficulties, it is possible to improve subjects' strategies. Before discussing this idea we should clarify the concept of strategy employed. Along with Van Dijk & Kintsch (1983) we believe that strategies imply the optimum use of a series of actions that lead to the achievement of a goal, thus reading comprehension strategies are an organised set of deliberate actions that the reader uses to achieve a specific goal leading to better comprehension. Among the strategies that have been taught we can highlight the identification of main ideas and summarising as having a basic role in the formation of a precisely detailed representation of the text, since they coincide with the construction of its macrostructure. As shown by Brown and collaborators (Brown & Smiley, 1977, 1978) the ability to identify the different levels of the ideas in a text cannot be taken for granted, but has to be considered as the fruit of an acquisition process throughout adolescence, generally achieved only after age 17 (approximately 12th grade). In accordance with their inefficient identification of main ideas, younger children tend to passively reread instead of underlining or taking notes. Other studies (Brown & Day, 1983; Brown, Day & Jones, 1983) found that subjects between the ages of 10 and 12, faced with the task of summarising, tended to employ a partially incorrect and passive strategy called copy-delete. This basically consists of selecting and copying, more or less verbatim, some ideas. In other words, the subjects neither applied thoroughly Kintsch and Van Dijk's macrorules nor constructed a summary based on a real interaction with the text, but "got the job done" inefficiently. The developmental pattern of these reading comprehension strategy studies can be described following Gardner (1987) who maintains that the progression leads from the absence of strategies to their flexible and efficient use with an inter-mediate stage in which the subjects are only able to employ the strategy partially, inflexibly and inefficiently.

Bauman's (1984) studies of main idea identification strategies and Hare & Borchard's (1984) studies of summarising are good examples of how both can be taught with positive results. These studies, along with more recent ones, employ the direct instruction method which simply consists of a complete and detailed explanation of the strategy in question (Winograd & Hare, 1988). The use of direct instruction coincides with the need for the subject to commit himself to apply the strategies intelligently. He/she must use them in an active, flexible and controlled manner, adapting them to different contexts and purposes. The teaching of strategies should not be limited to a detailed explana-tion of the behaviour and activities needed to carry them out along with practice, but should also try to intervene in those metacognitive aspects that lead to internal control and contribute to the subjects' disposition to actively employ them.

We have, to date, tried to emphasise the active nature of the strategies because we believe that this is the key to the difficulties in teaching them. We must draw a distinction between the subjects' knowledge of strategies and their real and effective use of them (Brown *et al.*, 1983). This is precisely where the problems arise. The important role of the subjects' disposition has been noted by several authors, especially Rothkopf (1988) who maintains that academic results are not dependent solely on cognitive competence but also on the disposition to employ them. Resnick (1987) goes even further when she states that the use of strategies, like any other higher thought process, demands effort on the part of the individual, and so it is necessary to cultivate not only the strategies but the disposition to use them.

Experiment

Objectives and hypothesis

Our objectives were the following:
(a) To design and test an intervention programme focused on developing an active disposition to comprehend and on the direct teaching of main idea identification strategies and outlining (structural summarising).
b) To test the effectiveness of Advance Outlines on structural recall and the interaction with the intervention on the subjects. To this end, in addition to the Outline group, we added a No Outline group and an Outline Construction group in which the subjects were simply asked to construct their own outlines.
c) To contrast different propositional analyses and measures of free-recall we used Kintsch's (Bovair & Kieras, 1985), as well as a more molar, less demanding system, employed by us in earlier studies (García Madruga & Martín Cordero, 1987).

In order to describe our study as exactly as possible we have used the tetrahedral model assumed by Brown *et al.* (1983), in which learning processes are decomposed through the interaction of four factors: (1) activities performed by the learner; (2) his/her characteristics, knowledge and skills; (3) the nature and sequence of the material to be learned; and (4) the type of criterial task employed, free recall, cued recall, recognition, problem solving, etc. Based on these four factors our study attempted to intervene in all of them in the following way. First, the subjects had to be persuaded to commit themselves actively to the task employing their strategies and knowledge to the fullest in order to actively understand the text and construct an adequate macrostructure. Secondly, we tried to improve their knowledge and reading comprehension skills by providing two crucial strategies. Thirdly, we presented them with three types of material and learning tasks (aid conditions). These were texts with Outlines, without Outlines and with the requirement that the subjects construct the Outline themselves. Lastly, we only used free recall as a criterial task but with several different measures of the dependent variable, including rote recall of the text's propositions, recall of the macrostructure and a measure of the structural recall.

Our hypotheses were the following:
(1) Intervention will produce a general improvement in the subjects' recall that is irrespective of the aids used. This improvement will be more pronounced in the measures of macrostructure and structural recall, and, specifically, in the Outline group.
(2) Each aid will produce a different effect on each type of measure. Given the characteristics of the subjects (third year high school) and of the experimental texts (relatively brief and simple), the No Outline group will score highest in propositional recall. The Outline group will score highest in the macrostructural and structural recall and the Construction group will score lower than the other two in the propositional rote recall.

Method

Subjects and design
The subjects were 90 third-year students from a lower-middle class high school in Madrid (mean age: 17.15 years). The TEA-3 verbal intelligence test confirmed the absence of special subjects and the homogeneity of the groups. One third of the subjects gave up or were unable to attend some of the sessions. The final number of students was 59 (No Outline: 20; Outline: 20; and Outline Construct: 19).

We employed a three random experimental groups design with pre- and post treatment measures. The levels of the independent variable were determined by the different aid conditions, which were randomly assigned to the subjects in each classroom. All the subjects received the same intervention. Control of the differentials was carried out based on the increase obtained. Since the comparison of the different measures of the dependent variable was an important part of the study, three measures were performed. Two of these measures were based on Kintsch's propositional analysis system. They consisted of the number of propositions and the number of macrostructural propositions recalled. We also performed another measure of the structural recall based on the number of main ideas recalled in the correct scenario plus the number of scenarios adequately recalled.

Materials

In the pre- and posttests the subjects used a Spanish version of the well-known texts, "Railroad" and "Supertankers". In the pretest, half the subjects received one text and the other half the other text. The texts were switched in the posttest. Both texts were signalled and contained 434 and 451 words respectively, with 10 hierarchical levels. Each text had 3 scenarios. In "Railroad" they were a representation of the theme 'groups in favour and opposed', and in "Supertankers" they were a statement of the problem: causes of sinking and possible solutions. As to the rhetorical structure, "Railroad" uses an adversarial, for and against comparison and "Supertankers" has a problem-solution organisation. Both texts are of equivalent difficulty. In each condition subjects received a booklet with instructions to read and assimilate the enclosed text that they would later be asked to recall. The Outline group also received an outline and was encouraged to use it while reading and assimilating the text. The booklet for the Outline Construction group asked the subjects to elaborate their own outline on the page provided. The answer booklets were the same for all subjects and instructed them to write everything they could recall as exactly as possible.

Table 1: The contents of the four sessions of the training programme

The first session consisted of:
(a) A general explanation based on:
  - What do you remember from the text? Main idea recognition skills.
  - What does the subject bring to his/her comprehension? An active process, knowledge and strategies.
  - What we are going to do?
  - Why is it useful?
(b) Presentation of text 1.
  - Suggesting titles for the text, discussion.
  - List main ideas, discussion.
(c) Modelling text 1. Four rules for active comprehension:
  - Notice the signals and rhetorical markers.
  - Try to visualise what is described.
  - Ask the text questions.
  - Reread to improve understanding.
The second session:
(a) Presentation of text 2.
  - Suggesting titles for the text, discussion.
  - Guided practice (active comprehension).
  - Discussion.
  - Feedback. Partial modelling.
(b) Presentation of text 3 (same as above).
The third session:
(a) Explanation of outline:
  - The structure of the relationships between the main ideas (using as examples texts 1 and 2).
  - Outline: structural summary. Rules for outlining:
    - Eliminate unnecessary details.
    - Group the information.
    - Use topic sentences.
    - Revision: Improve the connections and introduce new elements.
(b) Apply the above to the list of main ideas from text 3.
  - Discussion and modelling.
(c) Outline of text 4.
  - Students construct outline, guided practice.
  - Discussion.
  - Feedback-partial modelling.
The fourth session:
a) Outline of text 5.
  - Students construct outline, guided practice.
  - Discussion.
  - Feedback-partial modelling.
(b) Explanation of how to use an outline in processing a text. Discussion. Recapitulation.

The materials used during the intervention consisted of 5 texts (2 brief ones and 3 more complex and longer ones) and 4 booklets in which the subjects were to do their daily tasks. They also included reminders of especially relevant points of the instruction. The texts were "Las Pirámides", "Los Cheyennes", "Los Nómadas", "La Lengua Galesa" and "El Descubrimiento de América". All of them were natural and easy to understand extracts from encyclopaedias and social science publications.

Procedure and training programme
In the pretest session all subjects took the verbal intelligence test and were then given one of the experimental texts. Each subject was randomly assigned to an experimental group. They had 11 minutes to work with the text, afterwards they had 3 minutes of informal, unrelated, conversation with the researchers and were then given their answer booklets and asked to write everything they could remember for 12 minutes.

The training was carried out by two of the researchers in four 50-minute sessions during the normal class schedule. All subjects received the same session on the same day. The contents of the four sessions is shown in Table 1.

The posttest session was carried out 10 days after the pretest and was conducted in exactly the same manner as the pretest, the subjects received the alternate experimental texts, the only difference was the exclusion of the verbal intelligence test.

Results

The results of the three measures of the dependent variable can be seen in Tables 2 and 3 and Figures 1, 2 and 3.

Propositional recall

The ANOVA aid condition by pre- and posttest showed that both factors were significant (Aid Condition: $F = 11.10$, $p = 0.000$. Pre/posttest: $F = 6.11$, $p = 0.017$). The interaction between both factors was not significant.

The effect of the aid condition was also tested with two ANOVAs in pre- and posttest that reached significance (Pretest: $F = 8.80$, $p = 0.0005$; Posttest: $F = 7.84$, $p = 0.001$). In the pretest the differences between the No Outline group and the others were significant. In the posttest the Outline Construction group proved to be significantly different from the other two (Scheffé tests: $p < 0.05$).

Table 2: Pretest and posttest mean results for the three groups

	Propositional recall		Macrostructural recall		Structural recall	
	pre	post	pre	post	pre	post
No Outline (N = 20)	55.75	57.50	17.20	20.60	7.85	10.45
Outline (N = 20)	39.65	51.90	13.75	26.25	8.15	11.40
Outline Construction (N = 19)	36.26	38.84	13.37	18.68	6.84	8.84

Table 3: Percentage of increase between pretest and posttest

	Propositional recall	Macrostructural recall	Structural recall
No Outline	3.14 %	19.77 %	33.12 %
Outline	30.89 %	90.91 %	39.87 %
Outline Construction	7.11 %	39.71 %	29.23 %

Regarding pre- and posttest differences only the increase achieved by the Outline group reached significance (T-test: $t = -3.45$; $p = 0.003$). The use of non-parametric tests on the differential pre- and posttest scores in each group closely approached significance (Kruskal-Wallis: chi-square = 5,95; $p = 0.051$). The Outline group increase was significantly higher than the No Outline group's (Mann-Whitney: z=-2.29; p=0.022), approaching significance in the case of the Construction group (Mann-Whitney: z=-1.76; p=0.079).

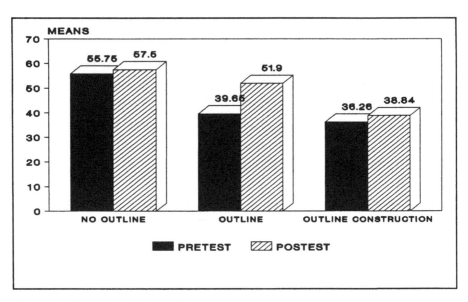

Figure 1: Propositional recall

Macrostructural recall

The ANOVA aid condition by pre- and posttest was only significant in the pre- and posttest factor (Pre/posttest: F=15,87, p=0.000). The ANOVAS performed to test the effects of the aid condition were not significant, with no significant differences between the groups in the pre- or the posttest.

The only difference in the pre- and posttest that achieved significance was the increase reached by the Outline group (T-test: t=-3.81; p=0.001), although the Construction group approached significance (T-test: t=-2.02; p=0.059). None of the differential scores between the pre- and posttest for any group achieved significance.

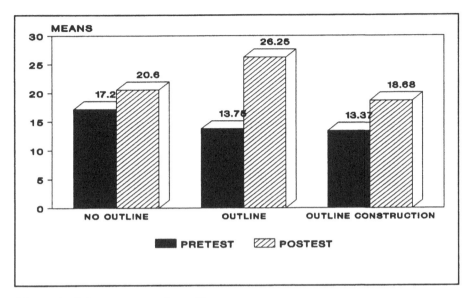

Figure 2: Macrostructural recall

Structural recall

The ANOVA aid condition by pre- and posttest was significant for both factors (Aid condition: $F=4.88$, $p=0.011$; Pre/posttest: $F=22.2$, $p=0.000$). The interaction between the two did not reach significance.

The ANOVAs performed to test the effects of the aid condition were significant only in the posttest ($F=4.54$, $p=0.014$). The difference between the Outline group and the Construction group also reached significance in the posttest (Scheffé test: $p<0.05$).

The increase obtained on the pre- and posttest achieved significance in the Outline group (T-test: $p=0.004$), and in the No Outline group (T-test: $t=-2.96$; $p=0.008$). The Construction group's results did not achieve significance (T-test: $t=-1.99$; $p=0.062$). None of the differential scores between the pre- and posttest for any group achieved significance.

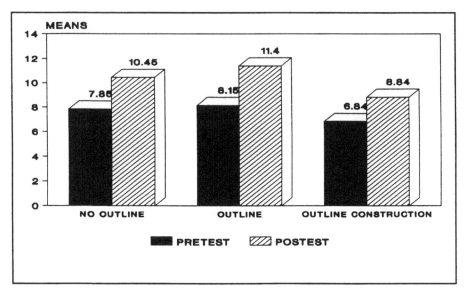

Figure 3: Structural recall

Discussion

In general the results seem to confirm our hypotheses. The Training Programme, in spite of its brevity, appears to have caused an improvement in the subjects' recall according to all three measures. Nonetheless, the differences between the pre- and posttest on the Propositions Recall achieved significance only in the Outline group. On Macrostructure Recall besides achieving significance in the Outline group, the Construction group approached it. The influence of the intervention seems to have been more general in the Structural Recall, reaching significance in both the Outline and No Outline groups, and also approaching it in the Construction group. It should be noted that the training produced some improvements in all three groups as shown in Table 3. Given the characteristics of the design, intrasubject control and the lack of a control group, the results must be confirmed by new studies that do not suffer from this limitation. Furthermore, it is imperative to consider basic issues related to possible long-term effects of the treatment and its transferability to other types of texts and situations.

The No Outline group's superiority in the "repetitive" propositional recall confirms our second hypothesis. As was to be expected, this
superiority disappears in the posttest when compared to the Outline
group. The results are less clear in the case of the Macrostructure: there
were no significant differences between groups to be found, in spite of
the expectation that there would be a difference in the Outline group's
favour. The cause of this last effect could be twofold. Besides the
experimental texts being brief, signalled and relatively simple, this could
be due to the fact that the measure of the macrostructure is a subset of
the measure of the total number of propositions recalled. In other
words, the No Outline group may have, in the worst case, learned large
portions of the texts, including the macrostructure, repetitively. No
differences can be observed in the Structural Recall of the pretest,
although the data point in the foreseen direction. In the posttest the
results are as expected, save for the fact that the Construction group
scored significantly lower than the Outline group. This was not
expected. The poor results of the Construction group in this last
measure, in the posttest, could lead us to ask ourselves whether the
training has really been effective for this group. We, are convinced,
however, that these data can be explained by the fact that the Construction group's task should not be expected to result in an increase on free
recall measures. On the other hand, this group's scores improved
between the pretest and the posttest on all the measures, though never
reaching significance.

The results appear to show, just as we expected, the complex
interaction of the aids, the training and the diverse measures of the
dependent variable. The effectiveness of our training in active strategies
seems to have been especially useful to the Outline group. This can be
explained on two levels. The use of advance outlines, as well as the
training, spring from the same cognitive theoretical frame of comprehension processes and the central role of the subject's prior knowledge
and strategies. It is therefore not surprising to find a convergence
between both types of intervention. A second, less optimistic analysis,
would highlight the fact that the training includes, precisely, teaching the
subjects to use outlines correctly. In any case, as stated earlier (Martín

Cordero, García Madruga, Luque Vilaseca & Santamaría Moreno, 1991) we believe that only with an adequate understanding of the cognitive processes involved in comprehension and of the active role that subjects must take it is possible to design effective aids and intervention programmes.

The results of the different measures bring several aspects to the fore that we would like to highlight. In the first place, we have shown that the aids, and to a lesser extent, the training, produce different effects on the diverse measures of recall. This fact underlines the need to take this into account in future studies by making qualitative as well as quantitative predictions (Mayer, 1988). This is being done to an increasing extent (see, for example, Cook & Mayer, 1989; Mannes & Kintsch, 1987). In the second place it enables us to evaluate the diverse types of measurement and propositional analysis employed, leading to a positive assessment of molar scoring systems, as well as the need to take into account the organisation of recall. Finally, the absence of results in the Construction group underscores the interest of including in future studies measures that go beyond mere free recall, since as Kintsch (1986) has stated, reading a text and learning from a text are two different things.

References

Ausubel, D.P. (1978). In defense of advance organizers: A reply to my critics. *Review of Educational Research, 48*, 251-257.

Barnes, B.R., & Clawson, E.U. (1975). Do advance organizers facilitate learning? Recommendations for further research based on an analysis of 32 studies. *Review of Educational Research, 45*, 637-659.

Bauman, J. (1984). The effectiveness of a direct instruction paradigm for teaching main idea comprehension. *Reading Research Quarterly, 20*, 93-115.

Bovair, S., & Kieras, D.E. (1985). A guide to propositional analysis for research on technical prose. In B.K. Britton & J.B. Black (Eds.), *Understanding Expository Text. A Theoretical and Practical Handbook for Analyzing Explanatory Text*. Hillsdale, NJ: Lawrence Erlbaum Associates.

Bransford, L.W., & McCarrell, N. (1975). A sketch of a cognitive approach to comprehension: Some thoughts about understanding what it means to comprehend. In W.B. Weimer & D.S. Palermo (Eds.), *Cognition and the Symbolic Processes*. Hillsdale, NJ: Lawrence Erlbaum Associates.

Brown, A.L., & Day, J.D. (1983). Macrorules for summarizing texts: The development of expertise. *Journal of Verbal Learning and Verbal Behavior, 22*, 1-14.

Brown, A.L., Day, J.D., & Jones, R.S. (1983). The development of plans for summarizing texts. *Child Development, 54*, 968-979.

Brown, A.L. *et al.* (1983). Learning, remembering, and understanding. In J.H. Flavell & E.M. Markman (Eds.), *Handbook of Child Psychology (Vol. 3)* (pp. 77-166). New York: Wiley.

Brown, A.L., & Smiley, S.S. (1977). Rating the importance of structural units of prose passages: A problem of metacognitive development. *Child Development, 48*, 1-8.

Brown, A.L., & Smiley, S.S. (1978). The development of strategies for studying texts. *Child Development, 49*, 1076-1088.

Carpenter, P.A., & Just, M.A. (1987). *The Psychology of Reading and Language Comprehension*. Newton, MA: Allyn & Bacon.

Cook, L.K., & Mayer, R.E. (1988). Teaching readers about the structure of scientific text. *Journal of Educational Psychology, 80*(4), 448-456.

Van Dijk, T.A., & Kintsch, W. (1983) *Strategies of Discourse Comprehension*. New York: Academic Press.

García Madruga, J.A., & Martín Cordero, J.I. (1987). *Aprendizaje, Comprensión y Retención de textos*. Madrid: UNED.

García Madruga, J.A., Luque, J., & Martín, J. (1989). Aprendizaje, memoria y comprensión de textos expositivos. Dos estudios de intervención sobre el texto. *Infancia y Aprendizaje, 48*, 25-44.

García Madruga, J.A., & León, J.A. (in press). Comprensión y memoria de textos expositivos. In M. Carretero & J.A. García Madruga (Eds.), *Adolescencia y Aprendizaje*. Barcelona: Laia.

Garner, R. (1987). Strategies for reading and studying expository text. *Educational Psychology, 22(3 & 4)*, 299-312.

Hare, V.C., & Brochardt, K.M. (1984). Direct instruction in summarization skills. *Reading Research Quarterly, 20*, 62-78.

Johnson-Laird, P.N. (1983). *Mental Models. Towards a Cognitive Science on Language, Inference, and Consciousness*. Cambridge: Cambridge University Press.

Kintsch, W. (1986). Learning from text. *Cognition and Instruction, 3*, 87-108.

Kloster, A.M., & Winne, P.H. (1989). The effects of different types of organizers on students' learning from text. *Journal of Educational Psychology, 81*(1), 9-15.

Mannes, S.M., & Kintsch, W. (1987). Knowledge organization and text organization. *Cognition and Instruction, 4*(2), 91-115.

Martín Cordero, J.I., García Madruga, J.A., Luque, J.L., & Santamaría, C. (1991). Improving learning and recall from text in distance education: Some experimental results. In M. Carretero, M. Pope, P.R.J. Simons & J. Pozo (Eds.), *Learning and Instruction. Vol 3*. Oxford: Pergamon Press.

Mayer, R.E. (1979). Can advance organizers influence meaningful learning? *Review of Educational Research, 49*, 371-383.

Mayer, R.E. (1983). Can you repeat that? Qualitative effects of repetition and advance organizers on learning from science prose. *Journal of Educational Psychology, 75(1)*, 40-79.

Mayer, R.E. (1988). Learning strategies: An overview. In C.E. Weinstein, E.T. Goetz & P.A. Alexander (Eds.), *Learning and Study Strategies*. Orlando, FL: Academic Press.

Resnick, L.B. (1984). Comprehending and learning: Implications for a cognitive theory of instruction. In H. Mandl, N.L. Stein & T. Trabasso (Eds.), *Learning and Comprehension of Text*. Hillsdale, NJ: Lawrence Erlbaum Associates.

Resnick, L.B. (1987). Instruction and the cultivation of thinking. In E. De Corte *et al.* (Eds.), *Learning and Instruction*. Oxford: Pergamon Press.

Rothkopf, E.Z. (1988). Perspectives on study skills training in a realistic instructional economy. In C.E. Weinstein, E.T. Goetz & P.A. Alexander (Eds.), *Learning and Study Strategies*. Orlando, FL: Academic Press.

Rumelhart, D.E. (1980). Schemata: The buildings blocks of cognition. In R.J. Spiro (Ed.), *Theoretical Issues in Reading and Comprehension*. Hillsdale, NJ: Lawrence Erlbaum Associates.

Winograd, P., & Hare, V.C. (1988). Reading comprehension strategies. In C.E. Weinstein, E.T. Goetz & P.A. Alexander (Eds.), *Learning and Study Strategies*. Orlando, FL: Academic Press.

13

The influence of an educator's interactive style on the process of comprehension in preschool-age children

Ottavia Albanese and Carla Antoniotti
University of Milano (Italy)

Abstract

A study of the language style used by adults in telling stories to nursery school children revealed two distinct interactive styles: (a) one-way story telling only, and (b) two-way dialogue based on the different opportunities given to the active involvement of the young child (Barbieri et al., 1987).

We had already studied the comprehension process of preschool-age children with regard to the intrinsic structure of a particular story (Albanese et al., 1985, 1986, 1988), and we considered it important in the light of these other studies to study the role that adult-child interaction plays in the process of comprehension. In particular, we hypothesised that the interactive style of the educator can influence the understanding of the story by the preschool child.

Introduction

Many recent studies have pointed out the importance of adult influence on child language development (Snow, 1972). In particular, the nature of adult-child conversation – especially how much opportunity the adult gives the child to function as a partner in communication – seems to play a large part in the child's linguistic formation; and the importance of the adult's guiding role in determining the nature of adult-child conversation has been verified both before and after the child has reached the level of verbal competence (Snow, 1977).

This influence is apparently sensed by mothers who read stories to their children, since they vary their style of interaction according to the age and the developmental level of the child (Ninio & Bruner, 1978; Snow & Goldfield, 1981-83; Teale, 1984): the mother adapts her pauses, questions, and information to the characteristics of the child's response. Rondal (1983) found the same phenomenon in kindergarten teachers who read stories to their students.

Another study of the language used by kindergarten teachers when telling stories to children (Barbieri, Devescovi & Bonardi, 1987) revealed two interactive styles: a narrative style, or one-way story-telling, and a dialogue style, or two-way story-telling. The two styles are characterised by different amounts of space left by the adult for the active participation of the child during the story-telling process.

No studies, however, had considered the effect of the teacher's interactive style on children's story comprehension. Having studied the process of comprehension in preschool-age children (Albanese & Antoniotti, 1985, 1986, 1988), we decided to study the influence of the teacher's interactive style on the process of children's comprehension. We hypothesised that the dialogue style of interaction – which leaves more space for the child's active participation – would lead to better comprehension of the story than the narrative style.

Method

The study was carried out in a kindergarten in Milan. Our sample was composed of 12 teachers and 24 children aged four and five. We selected the children carefully with a preliminary comprehension test, using the results to create a homogeneous sample. The story used was "The Wild Geese" from Alexandr Afanasev's collection of "Old Russian Tales".

Every educator told the tale to two children at a time. The pairs of children were chosen randomly from the sample established previously.

Soon after the story-telling, the comprehension of each child was evaluated using a set of 22 guided recall questions structured according to Stein and Glenn's (1979) "story grammar" categories: setting, initiating event, internal response, attempt, consequence, and reaction. The choice of this type of instrument to measure comprehension was based on findings by Trabasso *et al.* (1981) and by Brown (1975), and corroborated by the present authors (Albanese & Antoniotti, 1985), that free recall does not necessarily lead to the expression of all that the child has understood from the story, while guided recall by means of probe questions constitutes a useful method of obtaining more information regarding what the children have understood of the story (Stein & Glenn, 1979).

Every session was videotaped and transcribed in order to measure each child's comprehension and to evaluate the interactive styles of the educators. The children's comprehension was evaluated by classifying their answers to the set of guided recall questions as "yes" (correct) answers and "no" (incorrect or missing) answers. To identify the interactive style of the teacher, we adopted the two types suggested by Barbieri *et al.* (1987), narrative and dialogue style, which they defined by four parameters: (a) reconstruction of the story; (b) adaptation to the children's cognitive level; (c) cognitive checks on children; and (d) participation of the children.

From our observations of teacher and pupil behaviour during the session, we chose the aspects – e.g., questions posed to the children, frequency of spontaneous explanations, etc. – which could be used in our study to describe each parameter more specifically (see Table 1). We counted the occurrence of each aspect, and noted that there was a difference in the distribution of the aspects, permitting the individuation of two groups. We defined as narrative the style of the group of teachers in whose interaction the total instances of five of the aspects – number of responses to children's questions, number of explanations on children's request, number of rhetorical questions, number of questions asked by children, and instances of spontaneous interaction by the children – was greater than the instances of the other aspects, i.e., number of questions asked by the teacher, number of spontaneous explanations, number of non-rhetorical questions, number of responses received from the children, and instances of solicited interaction on the part of the children. We of course defined as dialogue the style of teachers in whose interaction the configuration was the opposite.

Table 1: Parameters for evaluation of the educator's interactive style

---

A. *Reconstruction of the story*: evaluation of story-telling style.
  · Number of questions asked to the children
  · Number of responses to children's questions
B. *Adaptation to children's cognitive level*: evaluation of the explanations.
  · Number of explanations given on children's request
  · Number of explanations given spontaneously
C. *Cognitive checks on children*: evaluation of types of questions made to the children.
  · Number of rhetorical questions (RQ)
    - no answer expected
    - used to draw attention
  · Number of encyclopedic questions (EQ)
    - often followed by explanations
    - used to verify knowledge
  · Number of questions referring to oneself (OQ)
    - referring to the child's personal world
    - used to involve the children more
  · Number of questions that verify comprehension (VQ)
    - often followed by explanations
    - used to verify comprehension of story events
  · Number of request of inference (RI)
    - used to verify children's ability to infer elements of the tale from story events
  · Number of requests of opinion (RO)
    - used to involve the children with personal opinions.
D. *Participation of the children* (space left to children by educator): evaluation of children's contributions to the interaction.
  · Number of questions asked to the educator (Q)
  · Number of yes/no answers to educator's questions (A yes/no)
  · Number of instances of spontaneous interaction (IS)
  · Number of instances of interaction encouraged by the educator (IE)

Based on these indices, five teachers were characterised by the narrative style (NS) and seven by the dialogue style (DS). Ten children, therefore, had been told the story in the narrative style, and fourteen had heard it in the dialogue style. The children's responses to the guided recall questions were analysed in relation to their assignment to the NS or the DS group.

Results

The percentage of "yes" answers of the children of the DS group (73.1%) was higher than that of the children of the NS group (59.1%). The difference between "yes" and "no" answers in the DS group was statistically significant (ANOVA: F=21.52; p<.001), and the same difference within the NS group was not significant (see Figure 1). The chi-square analysis yielded a significant difference between the answers given by the DS group and those given by the NS group (X^2=11.35; p<.001).

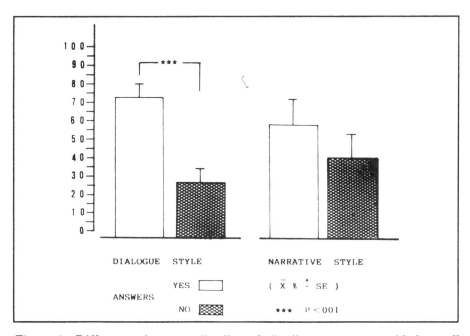

Figure 1: Difference between "yes" and "no" answers to guided recall questions in DS group and NS group children.

Regarding the story grammar categories, the differences between the two groups were relevant (ANOVA between percentage of "yes" answers in different categories: F=15.20; p< .001). More specifically, the children in the DS group responded more correctly in the categories

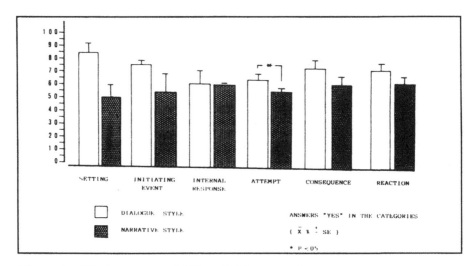

Figure 2: "Yes" answers of NS and DS groups by story grammar category

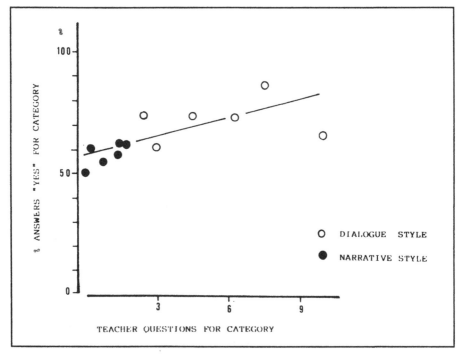

Figure 3: Correlation between the number of teacher questions during story-telling and "yes" answers to guided recall questions

"setting", "consequence", and "attempt", with the last category reaching statistical significance (67.8% vs. 57.5%; p<.05) (see Figure 2).

We analysed the relationships between each of the four individual parameters and the children's "yes" answers, and we discovered a statistically significant correlation between the total number of teacher questions during the story-telling (described in Parameter C, "Cognitive checks on children") and the percentage of "yes" answers of the children (r=.709; p<.001) (see Figure 3).

In addition, we noticed that three of the question types in Parameter C – Request of Inference (RI), Questions to Oneself (OQ), and Questions to Verify Comprehension (VQ) – were not even present in the storytelling of the NS teachers, but only in that of the DS teachers.

We also observed that the children in the DS group gave much more elaborate answers than the other children, i.e., linguistically richer and more elaborate answers.

Discussion

The criteria used for the individuation of the two interactive styles proved appropriate, since they allowed us to identify two styles with different characteristics and different results.

The dialogue style of interaction seems to be the more effective one, leading to better comprehension compared to the narrative style, especially in categories that other authors (Mandler & Johnson, 1977; Stein & Glenn, 1979; Levorato & De Zuani, 1984) have deemed the most difficult ones, such as the "attempt" category.

The correlation between the total number of teacher questions (parameter C) and the percentage of "yes" answers points to parameter C as a particularly important factor in influencing children's comprehension. In addition, the fact that questions of the RI, OQ, and VQ types were used only by DS teachers suggests their presence as a further indicator of the DS style. Since the "yes" answers were significantly higher in the DS group, the question arises whether these DS-specific aspects (the RI, OQ, and VQ question types) constitute the most important factor leading to children's comprehension.

In light of the present study, we intend to study in greater depth the dialogue interactive style, in order to determine which of its characteristics (with particular emphasis on the RI, OQ, and VQ question types) are more effective than others in leading to better comprehension.

References

Albanese, O., & Antoniotti, C. (1985). *Comprensione e memoria di una fiaba in soggetti in età prescolare*. Paper presented at the IV Congresso Nazionale Divisione di Ricerca di Base in Psicologia, Ravello, September.

Albanese, O., & Antoniotti, C. (1986). *Uno studio dei processi di comprensione di un testo in soggetti in età prescolare*. Paper presented at the V Congresso Nazionale Divisione di Ricerca di Base in Psicologia, San Pellegrino Terme, September.

Albanese, O., & Antoniotti, C. (1988). Il racconto della fiaba e la comprensione dei bambini. *Rassegna Italiana di Linguistica Applicata, 2*, 107-162.

Barbieri, M.S., Devescovi, A., & Bonardi, P.A. (1987). L'interazione verbale tra bambino ed educatrice durante il racconto di una storia. In S. Mantovani & S. Musatti (Eds.), *Adulti e Bambini: Educare e Comunicare*. Juvenilia.

Brown, A.L. (1975). Recognition, reconstruction, and recall of narrative sequences by preoperational children. *Child Development, 46*, 156-166.

Cochran-Smith, M. (1986). Reading to children: A model for understanding texts. In B. Schieffelling & P. Gilmore (Eds.), *The Acquisition of Literacy: Ethnographic Perspectives*. Norwood, NJ: Ablex.

Levorato, M.C., & De Zuani, E. (1984). *Il ricordo delle storie nei bambini: l'influenza della struttura del testo e del tipo di informazioni*. Università degli Studi di Verona.

Mandler, J.M., & Johnson, N.S. (1977). Remembrance of things parsed: Story structure and recall. *Cognitive Psychology, 9*, 111-191.

Ninio, A., & Bruner, J. (1978). The achievement and antecedents of labelling. *Journal of Child Language, 5*, 5-15.

Rondal, J.A. (1983). *L'interaction adulte-enfant et la construction du langage*. Bruxelles: Pierre Mardaga.

Snow, C.E. (1972). Mother's speech to children learning language. *Child Development, 43*, 549-565.

Snow, C.E. (1977). The development of conversation between mothers and babies. *Journal of Child Language, 4*, 1-22.

Snow, C.E., & Goldfield, B. (1981). Building stories: The emergence of information structures from conversation. In D. Tannen (Ed.), *Analyzing Discourse: Text and Talk, Georgetown University Round Table on Language and Linguistics*. Georgetown University Press.

Snow, C.E., & Goldfield, B. (1983). Turn the page please: Situation-specific language acquisition. *Journal of Child Language, 10*, 551-569.

Stein, N.L., & Glenn, C.G. (1979). An analysis of story comprehension in elementary school children. In R. Freedle (Ed.), *New Directions in Discursive Processing*. Norwood, NJ: Ablex.

Teale, W.H. (1984). Reading to children: Its significance for literacy development. In H. Goelman, A. Oberg & F. Smith (Eds.), *Awakening to Literacy*. Exeter, NH: Heinemann Educational Books.

Trabasso, T., Stein, N.L., & Johnson, N.S. (1981). Children's knowledge of events: A causal analysis of story structure. In G. Bower (Ed.), *Learning and Motivation* (Vol. XV). New York: Academic Press.